Risk Assessment
at Hazardous Waste Sites

Risk Assessment at Hazardous Waste Sites

F. A. Long, EDITOR
Cornell University

Glenn E. Schweitzer, EDITOR
U.S. Environmental Protection Agency

Based on a symposium

sponsored by the ACS Committee

on Environmental Improvement

at the 183rd Meeting of the

American Chemical Society,

Las Vegas, Nevada,

March–April 1982.

ACS SYMPOSIUM SERIES **204**

AMERICAN CHEMICAL SOCIETY
WASHINGTON, D. C.   1982

Library of Congress Cataloging in Publication Data

Risk assessment at hazardous waste sites.
  (ACS symposium series, ISSN 0097–6156; 204)

  Includes index.

  1. Hazardous waste facilities—Congresses. 2. Risk
—Congresses.
  I. Long, Franklin A., 1910–   . II. Schweitzer,
Glenn E., 1930–   . III. American Chemical
Society. Committee on Environmental Improvement.
IV. Series.

TD811.5.R57 1982   363.7'28   82–16376
ISBN 0–8412–0747–X   AACR2   ACSMC8   204
                                                    1–128   1982

ACS Symposium Series

M. Joan Comstock, *Series Editor*

FOREWORD

The ACS SYMPOSIUM SERIES was founded in 1974 to provide a medium for publishing symposia quickly in book form. The format of the Series parallels that of the continuing ADVANCES IN CHEMISTRY SERIES except that in order to save time the papers are not typeset but are reproduced as they are submitted by the authors in camera-ready form. Papers are reviewed under the supervision of the Editors with the assistance of the Series Advisory Board and are selected to maintain the integrity of the symposia; however, verbatim reproductions of previously published papers are not accepted. Both reviews and reports of research are acceptable since symposia may embrace both types of presentation.

CONTENTS

PREFACE

CHEMICAL WASTES resulting from the activities of U.S. industry and other groups have been dumped into waste sites around the country for many years. Typically the wastes are put into steel drums, not labeled, carried to an authorized dump site, and placed into a pit, which when filled is usually covered with earth. In some waste sites the filled containers are deposited on the surface. Bulk chemicals have been dumped in a few. Wastes other than chemicals are often deposited at the same sites, and illegal dumping has occasionally occurred, so there are undoubtedly unidentified sites.

Public concern over possible health and environmental impacts of chemical waste sites rose dramatically during the Love Canal episode and has continued to rise. Part of the reason for this concern is the growing awareness of the size and ubiquity of the problem of potentially hazardous waste sites. Another reason is increasing knowledge about the hazards that can arise from waste materials that once were viewed as virtually inert; a good example is PCBs (polychlorinated biphenyls). The extent of danger from older waste sites is particularly uncertain, partly because of the broadening list of potential hazards, and partly because of ignorance about the materials in most of the older sites. The current situation in the United States is that there appear to be many thousands of older dump sites that contain chemical wastes. Probably many, perhaps most, of these sites contain worrisome amounts of hazardous materials.

From the start, the Environmental Protection Agency (EPA) has been inescapably involved in this problem through the Resource Conservation and Recovery Act and earlier legislation, but since 1979 it became evident that EPA's normal resources were quite inadequate to deal with the overall problem in a timely way. Therefore in late 1980 the Comprehensive Environmental Compensation and Liability Act of 1980 (the Superfund Act) was passed. The Act revises a more limited National Contingency Plan to permit response to hazardous substances and to provide extra funds for EPA to tackle the problem.

A key difficulty in arriving at a rational approach to this many-faceted problem is establishing priorities, i.e., ranking waste sites in terms of potential hazards. This ranking presupposes a reasonably objective set of procedures to assess the risk offered by the sites, presuming that the most risky sites will be the first to be cleaned up (or rendered harmless). The question of risk is obviously a complex one, involving both health and environmental aspects, with the potential hazard depending on both the chemicals present and the ways hazardous chemicals might escape and impinge on people and the environment.

In March 1982 the American Chemical Society sponsored a symposium on risk assessments of hazardous chemical waste sites, and the chapters of this volume are the final versions of the papers that were presented and discussed at this symposium. The first chapters present the problem: the history of the development of Superfund legislation and the arguments about the most appropriate approaches to risk assessments, specific cases of hazardous waste problems in Louisiana, the problems of Love Canal and their bearing on risk assessment, and the impacts on human health that can result from hazardous waste sites. The next broad topic of the symposium was the central problem of methodology of risk assessment. The practical problems that confront the field teams who examine specific chemical waste sites are: what to monitor, how to monitor, and how to have reasonable assurance of the reliability of the results of monitoring. A final chapter considers a problem of central importance to the Superfund effort: how to incorporate risk assessment into the regulatory process.

The process of risk assessment for hazardous waste sites is still under development, and so necessarily is the development of regulations that build on risk assessment, but the commitment to utilize risk assessment appears firm.

The overall conclusion from the chapters of this book is heartening. The seriousness of the problem is clearly recognized, as is the need for risk assessment. Techniques and methodologies for assessment are fairly well in hand, although improvements in their application will surely occur. The social and political need for risk assessment is clearly recognized, and the incorporation of risk assessment into the regulatory process is well under way. The problems from hazardous chemical wastes are far from solved, but on the evidence of the contributions of this book, they are being attacked with intelligence and vigor.

FRANKLIN A. LONG
Cornell University
Ithaca, New York

August 3, 1982

Risk Assessment Under the Revised National Contingency Plan of Superfund

EDMUND B. FROST

Kirkland & Ellis, Washington, DC 20006

The 1980 Congressional revision of the National Contingency Plan, now known as the Superfund Bill, is effectively a compromise between sharply differing positions. It calls for priority setting to identify the most hazardous sites, and this in turn requires explicit assessment of the risks from the sites. Hence risk assessment criteria must be established so that there can be developed a national priority list of the hazardous sites that require remedial action. EPA has adopted the Mitre Model for assessing the risks and ranking waste sites. There is some criticism of this model, and there are a number of as yet unanswered questions about its application. Formal cost/benefit analysis will not be used, and risk assessment will be on a case-by-case basis. EPA will thus be developing its knowledge base and procedures as it carries out the program.

Legislative Background of Risk Assessment Under Superfund

The public and legislative debate which led up to the Enactment of the Comprehensive Environmental Response, Compensation, and Liability Act of 1980 ("Superfund") focused extensively on assessing the risks from old hazardous waste dump sites. Indeed, much of the controversy and difficulty in enacting the bill was a result of greatly differing perceptions about the risks posed by old dump sites and exposure to minimal levels of hazardous waste.

Those who took a fearful view of the risks thought that the cleanup program should eliminate all risk. On this premise they estimated 30,000 to 50,000 old sites requiring action and called for a massive cleanup program. Those who took a calmer view of the risks saw a serious but manageable problem and wanted a realistic assessment of the risks in order to determine the required extent of cleanup (1).

0097-6156/82/0204-0001$06.00/0

S. 1480, the original Bill reported by the Senate Environment and Public Works Committee (2) was a zero-release, zero-risk
bill. It contemplated clean-up of spills and old dump sites down
to a no-detectable-presence, zero-risk, level. No allowance was
made for risk analysis, comparative risk analysis, or balanced
decision-making where response action and its cost would bear a
reasonable relationship to the risk. Instead, cost factors were
put aside in favor of a liability scheme to tap deep corporate
pockets for whatever expense would arise.

H. R. 7020,(3) the House bill, was different from S. 1480.
It provided for prioritization of clean-up activities, and it provided for risk analysis and cost-effective clean-up.

By October, 1980, Superfund had bogged down in the Senate
over the extent of clean-up issues and liability issues. The
White House attempted to get the legislation moving again by
holding a series of meetings at the White House involving EPA,
the Justice Department, industry representatives and some of the
Senate Staff. Extent of clean-up was a major subject of these
discussions.

It emerged in the White House discussions that EPA was not
really interested in zero-risk clean-up. EPA just wanted to pre-
serve its discretion regarding level of clean-up and its leverage
over other responsible companies. Industry, on the other hand,
wanted to get on with reasonable clean-up, but did not want to
leave EPA and Justice with the discretion and leverage of S. 1480.
The fear was that they would insist on massively expensive and
unnecessary clean-up.

As the White House discussion wore on, an idea developed to
bridge between the polarized positions. The idea was that the
National Contingency Plan (NCP) for handling spills under Section
311 of the Clean Water Act could be expanded and tailored to provide a vehicle for making reasonable and prioritized decisions
about response actions and the level of clean-up. The revised NCP
idea was written up in several tentative drafts, but the White
House meetings and the White House effort broke up after Jimmy
Carter lost the 1980 election.

When the Lame Duck session of the 96th Congress opened in
late November, 1980, the Senate staff tried once more to move a
Superfund bill. This time they offered a compromise substitute
which used a revised National Contingency Plan as the center of an
integrated scheme to define a reasonable level of clean-up. This
substitute was ultimately enacted with few amendments.

Description of Key Superfund Provisions

Since it was hastily put together and received little analytical consideration, there are many technical flaws and there will
be much room for interpretation. Nevertheless, the bill as passed
contains an integrated set of response provisions centered on a

revised National Contingency Plan which calls for risk assessment and cost effectiveness.

Section 105 directs the President to revise the National Contingency Plan (NCP) (formerly limited to emergency response under Section 311 of the Clean Water Act) to include a new "hazardous substance response plan" setting out procedures and standards for response actions. This revision may be adopted only after notice and opportunity for public comment.

Among the elements which must be included in the revised NCP under this section are methods and criteria for determining the appropriate extent of either removal or remedial measures (Section 105(3)). The NCP must also contain means for assuring that "remedial" action measures, as distinct from the short-term "removal" measures, are "cost-effective" (Section 105(7)). Finally, Section 105(8)(A) requires criteria for determining priorities which are "based upon relative risk or danger to public health or welfare or the environment," and Section 105(8)(B) requires that such criteria be used to develop a national priority list of 400 sites.

Importantly, the penultimate sentence of Section 105 requires that, following publication of the revised NCP, "the response to ... hazardous substances releases shall, to the greatest extent possible, be in accordance with the provisions of the plan."

Section 104 of Superfund authorizes the President to remove or remedy any release or threatened release to the environment of any hazardous substance, as defined in Section 101(14) of the Act, or any other containment which may present an "imminent and substantial danger" to the public health or the environment. He may accomplish this by removing the hazardous substance, or by taking "any other response measure," unless he determined that such removal or remedy will be done by some responsible person. Regardless of which course the President chooses, Section 104(a)(1) states that it must be "consistent with the national contingency plan." Section 111(a) directs that the costs of such federal response be borne by the response fund created by the Act.

Section 106, entitled "Abatement Action," is the enforcement section. Under subsection (a), when the President determines that an actual or threatened release of a hazardous substance presents "an imminent and substantial endangerment to the public health or environment," he may direct the Attorney General to bring an action in a federal district court (against unspecified persons) to obtain "such relief as the public interest and equities of the case may require." The President also may take other unspecified action, including issuing such orders (to unspecified persons) as may be "necessary to protect public health and welfare and the environment."

Section 106(c) requires EPA, in consultation with the Attorney General, to establish and publish guidelines for using the response authorities under subsection (a). Unlike the revision of the NCP directed under Section 105, there is no express statutory requirement for public notice and comment prior to adoption of the guidelines. These guidelines are linked to the NCP by the direction in Section 106(c) that the guidelines "to the extent practicable be consistent with the natural hazardous substance response plan."

Application of the NCP to Private Clean-Up

The statutory provisions have failed to end the controversy over risk assessment. From the outset some environmental groups have argued that the risk assessment, cost-effectiveness provisions of the NCP are meant to apply only to government-financed clean-up under Section 104 and do not apply to private clean-up under Section 106 or to voluntary private clean-up (4). Instead, they advocated that private clean-up be handled on a separate track oriented towards Justice Department enforcement of complaints which would be based on a zero-exposure, zero-risk theory.

This interpretation of the NCP as being limited to government-financed clean-up is completely at odds with the statutory framework. Section 5 itself states its unqualified applicability to "the response to ... hazardous substance releases." Both Sections 104 and 106 involve the NCP as the reference point for response or clean-up. Sections 104, 105 and 106 all work together, with Section 105 giving the framework for determining priorities and extent of clean-up.

Further indication of the centrality of Section 105 of the NCP is found in Section 107. Section 107 outlines the only principles of liability included in Superfund. It states, without distinguishing between fund- and privately-financed clean-ups, that parties will be liable only for costs which are "consistent with" or in some cases "not inconsistent with" the NCP (Section 107(a)(4)(A)-(B)).

Although the legislative history is sketchy, Senator Stafford, one of the principal sponsors of the final substitute, recognized that private clean-up and orders for privately-financed clean-up under Section 106 would be governed by the NCP. In floor debate shortly before CERCLA's enactment, he described circumstances under which courts might review such orders. He stated that "we would expect the courts to examine the particular orders ... to determine whether they were proper, given the standards of the act and of the National Contingency Plan" (126 Cong. Rec. S. 15008 (daily ed. November 24, 1980), emphasis added).

As it stands now the EPA has made the NCP applicable to private clean-up as well as governmental clean-up in its Proposed Revisions to the National Contingency Plan (5). If the NCP is ultimately promulgated substantially as proposed, it is likely that the United States Court of Appeals for the District of Columbia will uphold it as within EPA's discretion. In the meantime attention of both industry and environmental groups has focused on the details of the Proposed Revisions.

Risk Assessment and Remedy Selection under the Proposed NCP

The proposed NCP recognizes the statutory mandate in Section 105(8) to develop risk assessment criteria and a National Priority List by limiting "remedial" actions to releases on the National Priority List (§300.67(a)). This will put teeth in the mandate of Section 105(8) and will assure the importance of the risk analysis/prioritization process.

EPA has not specified, however, that the National Priority List will be part of the National Contingency Plan, and there is still controversy over the procedure for making up the list and the function of the list.

The Mitre Model. With the assistance of the Mitre Corporation, EPA has developed a model for assessing the risk and ranking superfund sites by Section 108. The Mitre Model is a simple, straightforward model which will rank sites largely on the basis of easily obtainable information. Five pathways of exposure (ground water, surface water, air, direct contact, and explosion) will be considered. The probability and magnitude of an adverse event will be estimated for each pathway to arrive at an overall risk estimate.

Criticisms of the Mitre Model run from the fear that it will not be sufficiently precise or analytical to the concern that it makes no provision for utilizing extra information when it is readily available and is highly relevant.

It is beyond the scope of this paper to examine the Mitre Model in detail. Rather, it is sufficient to note that there is merit to EPA's decision to adopt a practical risk assessment/prioritization approach. If a full risk analysis had to be conducted at every site, EPA would spend all its resources on assessment and would have difficulty knowing where to start the clean-up. On the other hand, the model must be flexible enough to take into account the information which does exist regarding a particular site so that the risk assessment can be as accurate as practically possible.

Additional refinement of the Mitre Model is expected when the final NCP is published. In addition, EPA is expected to give some guidance in the final NCP as to how the ranking will work. Will EPA go strictly according to numerical priorities or work by groups or clusters of priority as suggested in the NCP? Will the states have a significant role in developing the priorities? What rights will private parties have to challenge the placement of their site on the priority list? All of these questions will have to be answered at some point.

Selection of Remedy. Risk assessment under the Section 105 scheme for the NCP is explicitly required only in connection with the prioritization process under Section 105(8). After prioritization on the basis of risk, the key issue is selection of remedy.

Section 105(3) requires "methods and criteria" for selection of remedy, and Section 105(7) requires that the NCP contain means for assessing that remedial measures are "cost effective."

There is an additional requirement when the President uses Superfund money for a remedial response. Section 104(c)(4) requires that the President "balance between the need for protection of public health and welfare and the environment at the facility under consideration and the availability of amounts from the fund."

The explicit cost/benefit balancing requirement of Section 104(c)(4) is not applicable to private response under the NCP. Thus, once risk is taken into account in the prioritization process, the EPA has considerable latitude in devising methods and criteria for remedy selection, with the major substantive requirement being cost effectiveness under Section 105(7). Nevertheless, in many cases the Mitre Model will not produce a risk assessment adequate for remedy selection, and a more detailed assessment will be necessary.

There has been much difference of opinion regarding appropriate methods and criteria for remedy selection. Annex XIII of the May 8, 1981 staff draft of the NCP called for clean-up to the level of Federal and State water quality standards, Federal water quality criteria, Federal drinking water standards, and national ambient air quality standards regardless of whether there was any substantial exposure or risk from the release to be remedied. The cost effectiveness requirement was applied only by calling for cost effective methods of reaching the standards imposed.

Industry criticized this approach on a number of grounds (6). Water quality criteria and national ambient air quality standards were never designed nor can they be sensibly used to set standards for individual releases. Water quality standards and Federal drinking water standards, although more appropriate for surface water or at-the-tap consideration, have been set for very few of the hazardous compounds which may be encountered at an old dumpsite.

Use of pre-existing standards and criteria to set mandatory levels of clean-up would preclude any site-by-site consideration of exposure and risk, and could result in huge clean-up costs where the risks were inconsequential. In addition, as a formal matter, the Federal water quality criteria for many substances have been left at zero pending full scientific and risk assessment, and use of these criteria would require a zero-release clean-up in a situation which imposed no significant risk in the first place (7).

The Administrator of EPA has taken the position that "each release presents a unique situation because of its diverse characteristics." (8) Based on this the option of requiring clean-up to federal and state standards and criteria was rejected:

"One option was to require cleanup to levels that
met Federal and State standards or water quality criteria. The Agency has decided that such a rigid requirement would impose the use of potentially inappropriate
levels of cleanup that would not allow consideration
of individual circumstances at each release. Any appropriate standard or criteria will be considered in determining the cleanup level of a particular release, along
with other technological and environmental factors." (9)

EPA also considered formal cost-benefit analysis as a decisionmaking tool and rejected it because it would "merely result
in rigid calculations that would foreclose any flexibility in the
ultimate decisionmaking." (10)
The decisionmaking method adopted in the Proposed Revisions
to the NCP is a case-by-case method where analysis and consideration must be given to a range of alternatives which have been
selected and analyzed. Cost, feasibility, adequacy, reliability,
and potential adverse impacts on health or the environment must
be analyzed. Any appropriate Federal and State standards and
criteria are to be considered. Finally, under proposed Section
300.67(j) the lead agency is "to choose the most cost effective
remedial alternative which effectively minimizes and mitigates the
danger and provides adequate protection of public health, welfare,
and the environment." (11)
The Environmental Defense Fund has objected to the lack of
standards criteria or goals in the Proposed NCP and considers the
proposed case-by-case approach too cost oriented. (12) Nonethe-
less, the proposed approach provides for assessment of the risks
to the public health and the environment, and consideration of
the impact of such risk is required in selecting the final alternative. Accordingly, risk assessment should be most important in
decisionmaking on the major Superfund sites. Professional experts
in this area will have a great challenge to develop their risk
assessment capacity and methodology on a case-by-case basis so
that it can indeed prove useful in selecting remedial alternatives
under the NCP.

Conclusion

In Superfund's Section 105 requirement for a Revised NCP,
Congress mandated a scheme for responding to releases of hazardous
materials. For old dump sites response starts with a survey and
prioritization of sites based on risk assessment. After prioritization the focus turns to selection of cost-effective remedies
which will adequately protect the public health and welfare and
the environment. In revising the NCP, EPA is engaged in a pio-
neering effort to develop a practical and effective site assessment model. After prioritization additional assessment will often

be required to select a remedy. Formal cost-benefit analysis will not be used, and the remedy selection risk assessment will be done on a case-by-case basis. In can be expected that EPA's experience in this case-by-case approach will be invaluable in developing future risk assessment methodology.

Literature Cited

1. Testimony of Dr. Louis Fernandez on behalf of Chemical Manufacturers Association, Hearings on S. 1480 before the Senate Finance Committee, 96th Congress, 2d Session, September 11, 1980.
2. Report No. 96-848, 96th Congress, 2d Session, July 11, 1980.
3. H. R. Report 96-1016, 96th Congress, 2d Session, May 16, 1980.
4. This double track approach was reflected in Annex XIII of the first, May 8, 1981, informal EPA draft of the Revised NCP. Section 300.2 of that annex specifically provided that response action would not be governed by the NCP.
5. 47 Fed. Reg. 10972-10995, March 12, 1982.
6. Chemical Manufacturers Association Comments on the Draft Revision to the National Contingency Plan, June 30, 1981, pp. 40-46.
7. CMA Comments on the Proposed Revisions to the National Contingency Plan, April 28, 1982, pp. 3-5.
8. Preamble to Proposed Revisions to the National Contingency Plan, 47 Fed. Reg. 10972, 10977, March 12, 1982.
9. Id., at 1078.
10. Ibid.
11. Id., at 10977.
12. Hazardous Waste News, May 17, 1982, 4, No. 20, pp. 154-155.

RECEIVED July 27, 1982.

Case Studies of Hazardous Waste Problems in Louisiana

WILLIAM B. DE VILLE

Louisiana Department of Natural Resources, Baton Rouge, LA 70804

The State's experience in dealing with hazardous waste management in Louisiana demonstrates the complexity of interactions among the technical, legal, economic and political elements of problems and actions. Current federal technical foundations for regulatory or "Superfund" decisions and actions provide only a rudimentary basis for program functions at the State level; this results in a tendency either to exaggerate or minimize the level of actual problems, and requires additional technical information to improve analysis of specific cases before action. This paper provides an analysis of the State's approaches to filling in gaps in the scientific and technical foundations for hazardous waste decisions by means of selected case histories. These case histories include (1) the approach to regulatory definitions of hazardous wastes, with emphasis on toxic properties; (2) the regulatory approach to standards and criteria for facilities managing hazardous wastes; and (3) development of cleanup responses to "abandoned" waste sites (with comments on the use of the 'Mitre Model' for ranking such sites). The paper supports the conclusions that (a) rational decisions can be made (though with some difficulty) in public policies related to the case histories, and (b) an agenda for improving the scientific and technical components of hazardous waste and "Superfund" decisions is now apparent and can be suggested in broad outline.

This paper presents the author's personal observations and conclusions derived from experience in hazardous waste regulatory program development at the state government level. Over the course of the past several years the State of Louisiana has been developing a very complex and comprehensive hazardous

0097-6156/82/0204-0009$06.00/0

waste regulatory program for protection of public health and the
environment. These regulatory efforts were also designed to
meet the goals of the Federal Resource Conservation and Recovery
Act for control of hazardous wastes; indeed, Louisiana was one
of the first states to receive interim authorization under the
Federal Act.

A major impetus to the passage in 1978 of the state
legislation authorizing the new hazardous waste program was the
recognition that Louisiana, as one of the centers of
petrochemical production in the nation, needed strict controls
over management of the inevitable wastes produced by such
industry. There were also past problems to be cleaned up. So a
two-fold effort was needed: (1) development of a new regulatory
program for prospective controls over future management of
hazardous wastes in the state; and (2) development of strategies
and programs to identify and resolve problems caused by poor
management practices in the past.

These efforts have been complicated, to a significant
degree, by the fact that they have been moving forward
concurrently with -- and often in advance of -- similar efforts
at the Federal level.

The inherent complexity of such a regulatory program

Particularly at a meeting such as this, which focuses
principally on scientific and technical matters, it must be
remembered that a regulatory program is a **legal device** to impose
certain constraints on activities in a specified segment of
society. The regulatory program is permissive as regards
certain activities which may be conducted within approved
boundaries, e.g., disposal of a waste in a facility that has
been permitted in accordance with the requirements and
procedures of the program. The regulatory program also forbids
other actions, e.g., "midnight dumping" of hazardous wastes, and
provides legal mechanisms and penalties for enforcing against
those actions that are prohibited.

The 'fit' of science and technology in the program.
Science and technology play important but, in fact, somewhat
secondary roles in the development of a regulatory program such
as that dealing with hazardous wastes. The first such role is
in problem definition: what are the characteristics that may be
used to identify those wastes for which regulatory controls are
mandated? The second general role calling for scientific
information and decisions is, what are the levels of risks below
which the regulatory system is permissive, i.e.,
standard-setting for allowed practices such as incineration or
land-filling of hazardous wastes. The third general role is in
measurement of parameters as stipulated by the regulatory
program to allow decisions to be made as to whether or not any

given hazardous waste management activity is allowed. Finally, there must be a chain of technical evaluations and decisions as to suitable technology and operating conditions for management of wastes in any given case, that will meet the constraints imposed by the standards and criteria of the regulatory program. Each of these roles places stringent demands on the ability of science and technology to supply information and answers.

I have characterized the scientific and technical components of the regulatory system as playing a secondary role in the regulatory development, because in fact they are not the driving force in development. Thus, it is entirely possible for the system to move forward in the face of partial scientific and technical input, or even, in the worst case, inadequate and unsatisfactory scientific and technical input and decisions.

In practice, the development of the regulatory system will tend to move forward regardless of the ability of science and technology to supply the various kinds of information demanded. Decisions must be made as to what waste materials are identified for regulatory control, what practices are acceptable, and what measurements are to be used for purposes of enforcement of the regulations. Perhaps the chief driving force resulting from the nature of the regulatory program as a legal 'machine' is toward legal clarity and simplicity of enforcement; this driving force can severely test the ability of the scientific and technical community to supply the information and answers needed to provide sound support to the regulatory program.

What is a hazardous waste -- and how do we test for it?

Louisiana's hazardous waste regulations were adopted and became effective in 1979, well in advance of promulgation of the Federal regulations. One of the critical problems we faced was defining just what it was that the regulations were supposed to control, with the additional concern that there must be standards of measurement or proof that could support enforcement actions against any violators of the regulations. Several possible courses of action were available.

Although the analogous Federal regulations for listing and identifying hazardous wastes were not yet in place, a set of Federal regulations had been proposed by EPA in December, 1978. In addition, several state hazardous waste programs then in existence had lists of 'hazardous wastes' in place (either by regulation or in practical use) as well as other means for characterizing or identifying wastes for control. We adopted as State regulations the wastes listed specifically or by source in the proposed EPA regulations. It was clear, however, that these listed wastes constituted only a subset of the wastes that should be put under regulatory control.

EPA had also proposed in December, 1978 a set of characteristics, together with criteria and test procedures, for

identifying wastes as hazardous. These were the characteristics
of ignitability, corrosivity, reactivity, and toxicity. This
scheme of identifying hazardous wastes based on characteristics
was adopted in the State regulations. We had little difficulty
in specifying parameters of these characteristics, e.g., flash
point or pH, together with test procedures, for the first three
characteristics. But the fourth, toxicity, was tied in the EPA
proposed regulations to an extraction procedure which was, we
concluded, not satisfactory for general identification of toxic
hazardous wastes.

A conservative approach to toxicity. We did not find a
simple technical approach to define the characteristic of
toxicity. Indeed, toxicity is not a simple property of matter,
but is a very complex relational property that is relative to
the living organism exposed, the specific material or
combination of substances at hand, the duration of exposure to
an organism, and the concentration of the substance. Adverse
impacts of a toxic substance on an organism may be manifested in
a short period of time (acute toxicity), or over a longer period
(chronic toxicity) that may involve many differing forms of
adverse impact, ranging from modification of behavior to
mutagenic or carcinogenic effects in affected organisms.
Because of the difficulty we experienced in finding simple
procedures to test for mutagenic or carcinogenic effects of
toxic substances in wastes, the approach in the Louisiana
regulations has been to a first-line criterion of acute
toxicity, based either on the presence of certain designated
materials in a waste stream, e.g., mercury, or on the listing of
components of the waste stream in standard and widely used
toxics registries or reference works. Thus, the definition of
the toxic characteristic was expanded far beyond the number of
substances specified by the EPA extraction procedure, and became
largely based on a comparison to literature references of the
constituents listed by an analysis of a waste stream (or
otherwise known to be present). Because the reference sources
to be used (e.g., Sax, NIOSH) tend to be quite conservative in
ranking relative toxicities, the distinctions between 'acute'
and 'chronic' toxicity tend to be blurred somewhat, so that a
great many substances with low acute toxicity but significant
chronic toxicity are included for regulatory control.
The next step was a regulatory requirement on those who
generated, treated, stored, or disposed of wastes to send in to
the State notifications identifying the wastes covered as
hazardous wastes by the State regulations. This notification
process tended to produce rather conservative identifications of
wastes; that is, notifiers tended to identify as hazardous
wastes a broader spectrum of wastes than have so far been
identified as hazardous by EPA. The 'universe' of hazardous
wastes regulated in Louisiana has been larger than that under

the Federal system from the beginning. However, a set of
procedures is available for delisting at the State level any
waste not identified as hazardous under the Federal system, by
demonstration that the waste (a) is not, on the basis of further
technical considerations, a hazardous waste under the State
approach, and (b) can be properly managed outside the hazardous
waste control system.

As a condition of interim authorization under the Federal
hazardous waste regulations, Louisiana has from time to time
formally adopted subsequent EPA listings of hazardous wastes.
It has been our experience that these waste streams have already
been covered and were already being regulated under the State
system. Intuitively, the State approach seems also to have
drawn upon the common sense of the regulated industries. Many
notifications of waste streams identified by the notifier as
meeting the toxic characteristic seem to have been made on
judgments as to the kinds of problems those wastes might cause
if not properly managed.

How big should the 'universe' of hazardous wastes be?

I must confess that, at a very early stage of regulation
drafting, we seriously considered the alternative of simply
calling all industrial wastes 'hazardous wastes,' because of the
difficulties of providing technical definitions of the term, and
specifying analytical procedures. We did not take that
approach, following further consideration of the intent of the
authorizing legislation: the approach to setting standards for
management of hazardous wastes called for more stringent, hence,
generally more expensive practices than would be required for
non-hazardous wastes. To regulate excessively wastes that
present little hazard would tend toward regulatory 'overkill;'
at the same time, failure to identify wastes requiring stringent
management would be regulatory 'underkill.'

Our experience in Louisiana is that the 'universe' of
hazardous wastes should be somewhat larger than that presently
defined by EPA, so as to accommodate certain waste streams
covered in the State program that, we believe, do need stringent
management, but are not currently addressed at the Federal
level.

The wastes themselves do not hold the answer. A
fundamental point is that a hazardous waste should be regarded
as only a **potential hazard**; if properly managed, it should not
cause any actual, significant adverse impact to public health or
the environment. The actual risks presented by a hazardous
waste are best defined in the context of how it is handled and
managed. Therefore, the key to the development of the
regulatory system became the selection of standards for the

management of hazardous wastes, designed to mitigate the hazardous nature of the waste.

This perspective is employed in all case-by-case decisions on petitions for exemption from the regulatory requirements, including petitions for delisting. The required technical information for evaluation of such petitions includes not only physical and chemical data on the waste material itself, but also information on how the waste would be managed if a favorable decision were made on the petition.

Setting regulatory standards and criteria

Still more difficult problems emerge at the level of establishing the boundaries of acceptable practices for managing hazardous wastes. These problems are in part technical, legal, and political.

Some technical problems. Implicit in the idea of a regulatory standard or criterion is some level of **risk assessment** and **risk evaluation,** more or less formally, and with more or less sophistication. By 'risk assessment' I mean that some sort of calculation or estimate is made, however roughly, as to the likelihood of an adverse impact of a hazardous waste on human health or the environment. Then 'risk evaluation' is some sort of calculation or estimate, however roughly, as to whether that likelihood or risk of adverse impact is acceptable, or not.

The Congress directed EPA (as the Louisiana Legislature directed the Department of Natural Resources) to promulgate regulations that establish minimum standards and criteria for acceptable management of hazardous waste. The regulation drafters must then seek scientific and technical information that enables risk assessments for various alternative practices in hazardous waste handling and management, particularly for development of any design or operations standards or criteria for hazardous waste facilities.

But at the same time the regulation drafters are faced with the still more difficult job of doing a risk evaluation. That is, they must recommend public policy decisions as to the acceptable levels of risks for a variety of hazardous waste activities, e.g., acceptable minimums for the efficiency of destruction of hazardous wastes by incineration. In principle, this process may be not unlike the decisions made by a banker about a potential loan, or an insurance underwriter in setting the rate for an insurance policy: a risk assessment is made, and an appropriate 'safety margin' is factored into the interest rate or the policy premium to arrive at an acceptable level of risk for the individual case (but based on experience and projections for a large number of cases).

But the regulation drafter has much less 'hard' information and experience available on which to base a risk evaluation, e.g., of acceptable levels of emissions or releases of a carcinogen into the environment. For reasons that will become apparent, therefore, there is a tendency to factor in larger safety margins into environmental regulations than most bankers would require on loans. There is a relatively short history of engineering data gathering and analysis for many waste management practices. But the real limiting problem is that of making evaluations based on potential health impacts of toxics at low levels. In the end, therefore, risk evaluations -- that is, standards and criteria -- tend to be largely based on policy considerations rather than on scientific and technical information per se. This situation will, one hopes, evolve to a greater role for scientific and technical information with the accumulation of experience and still more data and information. A reasonable goal for evolution of the regulatory program over time, therefore, should be improved **efficiency** of the program, i.e., allowance of least-cost practices that achieve the requirements of protection of public health and the environment.

It is apparent that science and technology should play a lead role for improving the efficiency of the regulatory program over time. There should be continuing evaluation of experience gathered in the operation of the regulatory system, together with provisions for incorporating new information from research and development efforts.

Some legal problems. The principle of **enforceability** plays a large role in the hazardous waste regulatory program. The regulation drafter, therefore, must pay a great deal of attention to measurement at the threshold of compliance/non-compliance with the regulations, and to mechanisms that will help build chains of evidence in cases of violation. As might be expected, a great deal of paperwork inevitably enters the regulatory requirements for these reasons. But there are also technical ramifications. The concept of enforceability tends to require the setting of specific quantitative measures for determination of compliance, as in evaluating groundwater monitoring well data. Among the technical complications in regulatory development is the need to set levels of quantitative measurements for enforcement purposes that reflect as well as possible the needs of program enforcement, and are also reasonably accurate and reproducible. As the need to monitor for low-level toxics is very great for a hazardous waste program, this has become a major technical area of difficulty.

The political problem. Paradoxically, the focus of public attention on the need to control and improve hazardous waste management decisions often makes those decisions still more

difficult than technical considerations alone might indicate.
While in many cases it might be argued from a scientific or
engineering perspective that a hazardous waste management
activity has met the requirements of the defined standards or
criteria, large sectors of the public may disagree. This may
well be true even in cases where all parties agree as to the
risk assessment, i.e., the likelihood of an adverse event or
impact. In the end, a risk evaluation is to a greater or lesser
degree a subjective decision. Perhaps the single greatest task
facing government in the implementation of the hazardous waste
program is the reduction of uncertainty and subjectivity at the
level of risk evaluation for hazardous waste management.

The Louisiana experience on standards and criteria

A heavy emphasis was given to use of **performance standards**
in the Louisiana hazardous waste regulations. Associated with
these performance standards are a number of design standards and
operating requirements designed to minimize releases or
emissions of hazardous wastes.
To a considerable degree, these regulations have satisfied
the technical and legal requirements and goals of the program.
Demonstrably, the program has resulted in great improvements of
waste management practices in the state. Certainly, no examples
of gross mismanagement of wastes, such as those which resulted
in some of the abandoned sites in the state, are now operating.
Whether the Louisiana program has yet satisfied the political
requirement, that is, satisfaction of the public's expectations,
remains in doubt -- as it does at the Federal level and in the
other states, as well.

Some mistakes of the past: "abandoned" sites

In early 1979, we had identified some 14 problem sites in
Louisiana that were the result of poor past management
practices. Typically, these sites represented grossly
unacceptable past practices, and not merely minor departures
from good engineering and operating conditions. Substantial
progress was made in getting several of these sites cleaned up
by responsible parties.
However, the effectiveness of state government to compel or
encourage private party cleanup of old problem sites appears to
have diminished at the same time public attention has been drawn
to these problem sites -- perhaps even because of that
attention. An additional factor which may be responsible for
slowdown in the rate of site cleanups seems to be the passage
and subsequent slow rate of implementation of Superfund.
Last year, the State proposed cleanup of an abandoned site,
following a feasibility investigation of the site that indicated
that on-site disposal of the wastes would be environmentally

acceptable, as well as the most cost-effective approach. Strong public reaction to the cleanup plan emerged, and no action has yet been taken. Aside from other technical problems and higher costs that would be associated with off-site disposal of the wastes from this site, another pattern is now visible and predictable; if disposal of these materials is proposed at any suitable commercial facility, this will arouse opposition in the local area of the recipient facility.

It seems unlikely that the public will not demand and encourage positive approaches to cleanup. Rather, the difficulty of building a consensus for proposed actions illustrates the complexity of the social and political issues that also play a part in governmental actions.

I suspect -- and hope -- that the slowdown is only temporary. It appears to have several causes. In addition to those already discussed, another seems to be the concerns of private parties about potential liability following cleanup, raised by the passage of the Comprehensive Emergency Response, Compensation and Liability Act (CERCLA, also known as Superfund). The question as to "how clean is clean" has become a question with not only technical but legal ramifications since the passage of CERCLA. As with the hazardous waste regulatory program, a combination of technical, legal and policy answers may be required to expedite resolution of existing old site problems.

The State has listed seven sites in the state for possible Superfund action . At one, several thousand drums of waste were stacked above ground, and then abandoned. Another represents a case of good intentions gone bad; a commercial waste site that lacked tight regulatory controls to set its operating conditions, and consequently got into trouble. Another started out as a municipal solid waste site, and somehow ended up as an uncontrolled dump for hazardous wastes. Still another was an old waste oil recovery plant that was operated very poorly, in a poor site area.

The 'Mitre Model' and site ranking. EPA has developed a hazard assessment model (commonly called the Mitre Model) to help rank problem hazardous waste sites for potential Superfund response. The model is a relatively simple one, with minimal data requirements. The model is designed to provide scores for actual or potential impacts on health or the environment by three routes of exposure -- groundwater, surface water, or air emissions -- as well as by direct contact or fire and explosion. Parameters include toxicity, quantity, physical state and persistence of the waste at a site; characteristics of a potential route (to groundwater, surface water, or air) of release of the wastes, e.g., soil permeability; the mode of exposure or use of the natural resource, e.g., an aquifer or surface waters used for drinking water; and the target of

exposure, e.g., population using the aquifer within a specified
radius of the site.

The Louisiana sites so far ranked using draft guidance for
data input to the model have not scored high, compared to many
other sites in the nation. In general, two factors account for
the relatively low scores: (1) low population density (except
in one case); and (2) subsurface geology and hydrology that
tends to minimize potential for groundwater contamination.

Our experience has led us to doubt the utility of the Mitre
Model for site ranking, beyond its use as a 'first cut'
screening device. The model scores for the Louisiana sites do
not appear to correlate particularly well with other means of
evaluating the relative ranking of problem seriousness, nor do
they give any useful insight as to the type or extent of
response that may be required.

In discussions with EPA, we have noted that the Mitre Model
does not appear to provide much ability to discriminate between
the relative levels of risks posed by different sites, nor does
it provide a basis for management decisions once a ranked list
has been gathered. On the other hand, it is clear -- again
based on our experience -- that the model would be too expensive
to use as a screening device at any level, if it were made much
more sophisticated than it is now. For example, the collection
and analysis of data on the 'groundwater route' can cost more
than $100,000. Louisiana has done rather extensive site
investigations for most of the sites listed with EPA, and the
investments of resources have been substantial. We believe
that, because of the simplicity of the Mitre Model, this poses
problems of comparability with other ranked sites, where much
less information may have been available.

Among the more serious problems in assessing an old site is
the difficulty and expense of characterizing the wastes. From a
technical point of view, it rarely makes sense to talk about a
'representative' sample, particularly if thousands of drums of
wastes from mixed sources are present.

RCRA and CERCLA present different technical problems. At
first glance, the technical needs for a hazardous waste
regulatory program, and for response to old problem sites might
seem nearly identical. Our experience indicates that this is
not so. The protocols for sampling and analysis at an operating
site can be well specified because the types and sources of
wastes are known, the management techniques and operations are
specified, and the regulated facility is required to develop
extensive records and an operating log. The objective of
sampling and waste analysis plans at an operating site is,
simply, quality control on rigidly specified and well documented
operations.

Quite different strategies for sampling and analysis are
required at an old problem site. In this case, little or no

documentation of the operations at the site -- including
chemical analyses of the wastes, management of the wastes, and
details and location of disposal areas -- is available. With
relatively limited resources, the responding governmental agency
must target its sampling and analysis efforts to meet the
objectives of problem definition (the hazardous materials
present, the extent and nature of contamination of the site, and
the surface and subsurface characteristics of the site); and
development of a feasible strategy to address the problems
(including, for example, an assessment of the suitablilty of the
site for on-site burial of part or all of the waste materials).
Inevitably, the available data will be limited by contrast to
the data that should now be available from a properly operated
hazardous waste management facility that is subject to the
current Louisiana or RCRA requirements. If a drum of waste at
the abandoned site is sampled, the governmental agency must bear
the cost of recontainerization or immediate disposal of the
wastes, and an average analytical cost that may range well
upwards of $500 per drum sampled. Given a site, for example,
where containers have already deteriorated, the concerns for
analysis will probably not center on whether the wastes are
incompatible (they have already reacted, if so), but on the
technical feasibility of on-site versus off-site (or some
combination) disposal of the wastes present.

Degree of hazard and level of control

One of the most promising technical approaches to
regulation of hazardous wastes (and one that already seems in
progress for evaluation and response to old problem sites) is
the development of a 'degree of hazard' system of waste
classification. The current RCRA approach and, to a slightly
lesser degree, the Louisiana approach tend to treat all
hazardous wastes as about equal. In fact, however, it is
apparent that waste materials differ very greatly over a
spectrum of **intensity** of the characteristics that make them
hazardous. This is particularly so for the characteristic of
toxicity.

The degree of hazard approach, on the face of it, may offer
a similar basis for prescribing a sort of intensity of need for
management. Beyond this, it is also apparent that, given the
choice of available management options for a given waste stream
(such as land treatment, land burial, incineration, chemical
fixation, etc.) there can be technical decisions rendered as to
which of these options are suitable for the waste material
(e.g., a chlorinated hydrocarbon solvent waste), and which are
less suitable, or even unacceptable.

The technical challenges to the development of a hazardous
waste management scheme based on such premises are great. For
example, standard methodologies for evaluating the probable

performance of technical options for treatment or disposal of a waste material, and improved and simplified methodologies for risk assessment, need to be developed.

But the potential usefulness of simplified management tools to choose the optimal level of control and technology for handling hazardous wastes may well be a driving force to future designs, not only of hazardous waste treatment facilities, but also of the regulatory system itself.

RECEIVED July 7, 1982.

Determining the Impacts on Human Health Attributable to Hazardous Waste Sites

VERNON N. HOUK

Center for Environmental Health, Centers for Disease Control, Atlanta, GA 30333

The Superfund Act and implementing Executive Order
assign to the Department of Health and Human Services
the responsibility for assessing the threat hazardous
waste sites pose to the health of workers and the
general public. Our working definition for such a
health risk assessment is the determination of proba-
bilities of various adverse health outcomes that
result or would result from exposure to specified
hazards. To fulfill our legal mandate, we intend
first to determine the nature and quantity of toxic
materials present at Superfund sites, the size and
proximity of potentially exposed human populations,
and the existence of likely pathways for significant
human exposure. If all three factors are present,
additional study through various approaches, such as
exposure documentation studies, disease-related
epidemiologic studies, or disease registries, will be
considered.

In December 1980, President Carter signed into law a bill
known as the Comprehensive Environmental Response, Compensation
and Liability Act--more commonly known as the Superfund Act. It
is designed to deal with the burgeoning problem of cleaning up
old and abandoned hazardous waste sites. The Act imposes a tax
on the chemical industry and then allocates that money, along
with a Federal and State contribution, to sites needing
emergency response or long-term cleanup; assessment of health
effects which may result from exposure; and enforcement action
against parties responsible for these sites, including generators, transporters, and site owners.

Officials of the Environmental Protection Agency (EPA)
testified before Congress that EPA had identified over 9,600
uncontrolled hazardous waste sites, had made preliminary
assessments of 6,100 of these, and, by July 1981, had completed

almost 2,800 on-site inspections, including some sampling of
site contents and ground and surface water.

The Centers for Disease Control (CDC), an agency of the
Public Health Service, has been assigned the responsibility for
implementing the health-related portions of Superfund for the
Department of Health and Human Services. Although CDC has the
principal role in measuring the impact or potential impact on
human health which may be related to exposure, the National
Toxicology Program and the National Library of Medicine have
active roles in providing up-to-the-minute information about the
known health effects of chemicals or chemical combinations found
in Superfund sites. Similarly, CDC's National Institute for
Occupational Safety and Health will help ensure that the health
of emergency personnel and cleanup workers is protected when
they respond in Superfund actions.

The tremendous number of orphaned waste sites in this
country reflects the growth of the synthetic organic chemical
industry (Figure 1). In 1940, approximately one billion pounds
of synthetic organic chemicals were produced in the United
States. By 1965, this quantity had increased nearly a hundred-
fold and it now exceeds 300 billion pounds a year.

This exponential growth of the industry has a worrisome by-
product in the massive problem of toxic waste disposal. To
determine if such wastes cause health damage in exposed human
populations, we must conduct epidemiologic investigations of
relationships between toxic exposure and possible adverse
health outcome, clinical or subclinical. In conducting human
health studies before Superfund was enacted, we found that such
investigations are conceptually simple and involve straightforward
concepts of cause and effect. Implementing these studies,
however, is "easier said than done."

Because of the extremely diverse situations in which
hazardous materials are involved, each study presents unique
problems. Table I illustrates the diversity, both in waste
materials and exposure settings, of eight recent toxic waste
situations. Differences, in fact, can be quite large, ranging
from the common drum-filled dump site to the widespread
dispersal of waste material. Some generalizations, however, can
be made about these kinds of studies.

Epidemiologic studies usually consist of three fundamental
phases: (1) Determining what toxic materials are present,
(2) establishing how human exposure to these toxic materials
might occur, and (3) measuring actual or potential biologic
effects. Substantive information in each of these phases is
essential for successfully completing any epidemiologic study.

The first step in any toxic waste study is to determine
what toxic materials are present and in what amounts. Since
most toxic waste situations involve the dumping of diverse
materials, problems of expensive technical methodology can be
formidable. However, without an adequate inventory of the

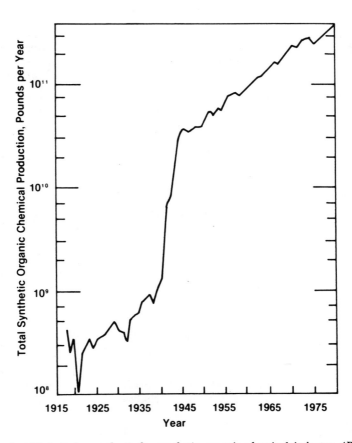

Figure 1.   Historical growth of the synthetic organic chemical industry. (Reproduced from Ref. 15.)

TABLE I

Recent Situations Involving Potential Human
Exposure to Potentially Toxic Waste Materials

Location of Site	Toxic Materials	Physical Condition	Principal Routes of Potential Human Exposure
Love Canal dump, Niagara Falls, New York	Largely hydro-carbon residues from pesticide production	Inactive land-fill in resi-dential area	Direct, air-borne, and waterborne contacts
Melvin Wade dump, Chester, Pennsylvania	Diverse organic chemicals	Surface col-lection waste in drums in urban setting	Direct contact, explosion, and fire
Woburn, Massachusetts	Arsenic com-pounds, heavy metals, organic chemicals	Abandoned waste lagoon with multiple surface dumps	Direct and waterborne contacts
Triana, Alabama	DDT and related compounds	Industrial waste dumped in a rural stream	Food chain (fish)
Bloomington, Indiana	Polychlorinated biphenyls (PCB's)	Industrial waste contami-nating munici-pal sewage used for garden manure	Direct contact and possibly food chain
Tristate Mining District, Oklahoma, Kansas, and Missouri	Heavy metals, acidic aquifer	Mine tailings and acidic aquifer recharge	Airborne and irreparably contaminated aquifer
Montgomery County, Pennsylvania	Trichlor-ethylene	Industrial waste contami-nating aquifer; underground storage tank rupture	Direct and waterborne contacts
Pittston, Pennsylvania	Chlorinated solvents, HCN, and heavy metals	Millions of gallons of waste dumped into abandoned mine shafts	Food chain and waterborne exposure

chemicals--their quality and quantity and the physical conditions
under which they are present--it is premature, if not impossible,
to design adequate epidemiologic studies. If only low concentrations or small amounts of toxins are present, there may not
be sufficient reason to proceed with investigations. The
decision of whether to proceed is not always simple, since
public concern and political pressure can be compelling even in
the absence of a confirmed toxic exposure.

Despite such pressures, it is important to realize that
conducting even a simple survey of health effects is perilous
if one lacks information about toxins and exposure, since few,
if any, measurable health effects are sufficiently specific for
exposure to particular toxins or groups of toxins to be surrogates for directly measuring toxic exposure.

After assessing the nature and quantities of toxins present,
one must evaluate their potential for human exposure. Even when
materials known to be toxic are present, human exposure may not
have occurred or may be only remotely possible. Before exposure
can be evaluated, the means of exposure (for example, direct
contact, contaminated water, or contaminated air) must be
determined, and the size of human populations potentially exposed
and their degree of proximity to the toxic materials must be
defined.

Another reason one must consider the nature of the chemicals
involved is that some chemicals pass through the body quickly,
whereas others are stored in tissue. Exposure to chemicals
which persist in tissue, such as Polychlorinated Biphenyls
(PCB's) or DDT, of course, provides much greater opportunity
for productive epidemiologic study than exposure to transient
agents (such as trichloroethylene in Montgomery County). Many
studies are undertaken long after active exposure has occurred.
In such studies, exposure can be judged objectively by measuring
levels of persistent toxins in tissue.

The prime objective in epidemiologic studies is to associate
particular exposures with potential health effects and thus to
define cause-effect relationships. Since this process is an
indirect assessment, it is highly dependent on the accuracy and
specificity of observations recorded both for exposure and
outcome. It is a more powerful study if dose-response relationships can be shown, that is, if increasing levels of exposure
are associated with increasing frequency of the health effects
in individuals.

To determine this cause-effect relationship, the epidemiologist must be aware of four difficulties which can limit the
power of an epidemiologic investigation (Table II). The first
involves the size of the population needed for a study to demonstrate a given health effect with a given degree of power. This
depends both on the degree of exposure and on the expected
baseline frequency of the particular health effect. If the

TABLE II

Epidemiologic Issues Fundamental to
Evaluating Potential Toxic Waste Health Effects

Epidemiologic Issues	Impact on Epidemiologic Study
1. Expected baseline frequency of specific health effects	Relatively low expected frequency requires large population for study, especially to detect small increases in risk.
2. Latency period	Long latency may require periodic or continuous long-term population followup.
3. Multiple causative factors (clinical nonspecificity)	Since particular health effects are not often specific for particular toxic exposures, data must be collected and evaluated.
4. Alteration of routinely collected exposure and/or outcome information to coincide	Diminishes precision of exposure/outcome relationship.

health outcome to be measured is rare, the population to be studied needs to be large.

The second difficulty is that of <u>long or variable latency,</u> the period between exposure and measurable effect. At lower levels of toxic exposure, outcomes such as cancer may not occur until years later. This means that the study design must include long-term health followup. The usual recommendation is to establish an exposure registry, with a 20- to 25-year study. The alternative is to study current cancer in a population known to have been exposed over a span of years. Neither alternative is easy since registries are expensive and complicated logistically by the mobility of the U.S. population.

The third difficulty is that of <u>competing causes</u> or, expressed differently, the clinical nonspecificity of the health effects under study. To the epidemiologist, this means adjusting for many potential confounding factors in a study. The study design must include collecting data on other exposures that might also cause the health effect. In practical terms, this means collecting data on past occupational exposures, personal exposures, such as cigarette smoking or the use of alcohol or drugs, and personal characteristics, such as sex, race, age, and socioeconomic status, which may predict levels of risk for specific disease states. The more variables a study addresses, of course, the more complex its eventual analysis and the greater the size of the population needed for adequately assessing health effects.

The fourth difficulty—one often encountered—is that a health outcome routinely measured by political boundaries (cancer mortality, for instance) has to be reshaped to <u>environmental boundaries</u>, such as those imposed by an aquifer or wind pattern, to include the population under study. Unless the exposure or outcome borders are defined by properly extrapolating or interpolating them to coincide geographically, any cause-and-effect relationship is useless.

Several methods the epidemiologist has for investigating relationships vary greatly in cost, time and energy expended, and analytic and interpretive value. The descriptive study is simply, as the name implies, a series of rates, ratios, and proportions which help describe either the exposure or the outcome in detail. The case-control study is relatively easy to carry out, and new analytic techniques have already increased its interpretive value. An exposure study, although not strictly in the realm of epidemiology, documents exposure in various substrata of a population.

The following three accounts of waste-site investigations demonstrate four major approaches to assessing health effects at hazardous waste sites: (1) Descriptive studies, (2) case-control studies, (3) studies, and (4) cohort studies.

Woburn, Massachusetts - An Example of Descriptive and Case-
Control Studies

     During the summer of 1979, residents in this eastern
Massachusetts town became concerned an apparent cluster of six
leukemia cases, diagnosed since 1969 in children of families
living in a 6-block area in the southeastern portion of the
town (1970 total population: 37,067). The cluster was reported
to the Massachusetts Department of Public Health (MDPH) and to
the Centers for Disease Control (CDC) both by local citizens and
a Boston physician. Concern was also expressed regarding other
cancers, especially kidney cancer. An assessment of town-specific
cancer mortality rates made independently by MDPH at about the
same time for the 1969-1978 decade in Massachusetts showed
statistically significant elevations in Woburn for all cancers,
as well as for several specific kinds of cancer.
     The concern about excess cancers focused on possible
causes related to toxic waste disposal sites in the town. From
the mid-19th century until the 1920's, Woburn had been a major
center first for commercial tanning of hides and later for
chemical production of lead arsenical pesticides. Efforts in
the 1970's to develop an industrial park in the northeastern
part of the town uncovered several old abandoned waste disposal
sites containing hides or chemicals related to prior industrial
activities. Among the toxic chemicals found in excessive amounts
were arsenic, lead, chromium, and cadmium. Concern about leaching
of waste chemicals into drinking water led to extensive testing
of wells which supply the town's water. Two of 11 such wells
showed excessive levels of various organic chemicals (trichloroethylene, for example). These two wells had supplied water,
principally to the eastern half of the town, since the mid-
1960's and were located a short distance to the north of the
leukemia cluster neighborhood. Because of the demonstrated
chemical levels, the wells were closed in 1979.
     To explore the possible cause of reported cancer excesses,
and particularly to examine the possible relationship between
the town's toxic waste situation and the childhood leukemia case
concentration, MDPH and CDC, in 1980, conducted a joint study.
Incidence patterns for selected types of cancer were studied by
using data collected from local and regional hospital sources.
Increased frequencies for childhood leukemia and for kidney
cancer were confirmed, with the childhood leukemia excess being
located in one particular census tract.
     Shifting epidemiologic gears to case-control methods,
investigators then obtained additional data regarding a wide
range of possible environmental causes through detailed interviews with patients or relatives of patients. For all 12 childhood leukemia cases in the town, this process included interviews
with two age- and sex-matched controls drawn from school enroll-
ment lists--one from the matched case neighborhood, the other

from another part of the town.   No causal factors were found
from these interview data for either kidney cancer or leukemia.
For leukemia in particular, no clear-cut differences were seen
in responses provided by case and control families.

The limitations of these observations can be described in
terms of the general problems associated with such toxic waste
studies.   The health outcome of concern (cancer, specifically
childhood leukemia) is a relatively rare disease.   The toxic
chemical exposure of particular persons having these diseases
could not be directly documented, since the organic chemicals
found in well water were transient.   Therefore, no means were
available for linking particular leukemia cases to particular
toxic substances.   For the present, then, we are left with an
apparent leukemia case cluster and no good evidence for
associating cases with the striking environmental toxic waste
problems clearly present in the town.

Triana, Alabama - An Example of a Cross-sectional Study

Triana, with a population of 600, is located at the
confluence of Indian Creek and the Tennessee River.   From 1947
until 1971 DDT was manufactured in a plant 10 kilometers from
Triana.   Several thousand tons of DDT industrial waste accumulated in the sediments of a tributary of Indian Creek.   Locally
caught fish have had total DDT residual levels up to 100 times
the tolerance of 5 parts per million (ppm) set by the Food and
Drug Administration.   The high levels of DDT residues in the
12 persons surveyed were suspected of resulting from their
consuming fish which had accumulated DDT from sediments.   The
task at hand was to measure residents' exposure to DDT and
relate it, if possible, to health effects in the population--an
example of a cross-sectional study.
    Of the 518 persons participating in the study, including
44 commercial fishermen and their families, 96.3 percent gave
blood specimens.   The mean serum DDT level (76.2 parts per
billion (ppb)) of 499 persons living downstream from the DDT
manufacturing plant was about four times the national mean
(16.7 ppb).
    The number of participants with high serum levels of DDT
was small, and therefore no sound conclusions could be drawn
from the study, regardless of the time involved.   On the
other hand, the population was large enough for the study to
increase understanding on several points--namely, that DDT
appears to accumulate with age, that the liver is affected, and
that serum triglycerides and cholesterol levels are related to
DDT level.   Further, in the citizens of Triana, levels of PCB's
in serum were found to correlate positively with blood pressure,
certain liver function tests, and cholesterol, independently of
age and sex.

Bucks and Montgomery County, Pennsylvania – An Example of a Potential Cohort Study

In May 1979, the Pennsylvania Department of Health reported that several public and private water supplies in Montgomery County, Pennsylvania (population 627,600) were contaminated with trichloroethylene (TCE). TCE was found in 70 of 100 well-water supplies in Rahns, Pennsylvania. Ten of these positive samples showed levels of TCE in excess of 500 ppb, and 5 samples showed TCE in excess of 1,000 ppb. The suspected source of contamination was a steel wire and rod mill that used this compound as a degreasing agent. Improper disposal resulted in spills on plant property, which leached into nearby lagoons and well water supplies. At another tube company, a pipe feeding a large storage tank of TCE ruptured and spilled over 2,000 gallons of TCE into a storm sewer, resulting in further contamination of surface and ground water.

Apparently, hundreds of people were being exposed to TCE through their drinking water. No acute illness had been reported to be related to the contaminated well water, but Congressional requests and a high degree of concern from citizens and local officials led to further study.

Seven (6%) of 117 residents screened for trichloro metabolites had detectable levels of trichloroethanol and trichloroacetic acid in their urine. The approach used in this investigation was to specifically compare deaths from liver cancer in Montgomery County with such deaths in unexposed populations within the same county as the primary comparison group and with populations represented by State and national data as alternate comparison groups. To a certain degree, this situation lends itself to a prospective cohort study. Had this epidemiologic tool been employed, a discrete population of exposed, and perhaps unexposed, individuals would have been registered, with appropriate identifiers, and assigned a particular exposure level. This population would then be actively followed for a minimum of 15 to 20 years, and vital status and any designated adverse health outcomes would be recorded and compared with the State or national rates for the same event. Obviously, a very long time is required and followup is very expensive.

However, information obtained from a prospective cohort study which is well designed and meticulously conducted carries tremendous weight when one is balancing information from less powerful investigative tools. These prospective cohort studies must be well thought out, and there should be a reasonable possibility for funding through the life of the study.

An alternative to the prospective cohort study is the retrospective cohort study, in which a defined population with a past exposure has an existing measurable health outcome. The past exposure needs to be well documented so that a given level of exposure can be assigned to each individual in the population.

This, of course, requires accurate and accessible exposure
records which can be readily coupled to the population being
viewed retrospectively.

Thus, we have three situations in which four different
epidemiologic approaches could be used to investigate health
effects associated with hazardous waste sites or any environmental contamination.

In summary, with the increasing public concern about toxic
wastes growing out of such incidents as Love Canal, it is
incumbent upon the Federal Government to try to answer the
public's question: "How is my health, and that of my family,
affected by exposure to toxic wastes?"

If I can leave you with only one message, it is that this
question is extremely difficult to answer.

As I have shown the epidemiologist or the investigator can
take any one or several of various approaches to the study of
possible health effects from exposure to toxic substances at a
given dump site or chemical spill. Each approach has its
strengths and weaknesses. Similarly, each dump site or spill
has its peculiarities. We are in the process of matching what is
known about the priority sites to be cleaned up under the
Superfund law with the best approach to learning about the
possible health effects on exposed people living around those
sites. Once we have sorted out the sites most likely to give us
answers about health effects and have designed the best research
strategy for studying each site, we can begin our work in the
field. This work promises to be difficult and complex, but we
hope it will resolve these public health concerns.

Literature Cited

1.   American Conference of Governmental Industrial Hygienists
     (ACGIH). "Threshold limit values for chemical substances and
     physical agents in the workroom environment with intended
     changes for 1980." Publication Office, ACGIH, PO Box 1937,
     Cincinnati, OH 45201.
2.   Cannon, S.B.; Veazey, J.M.; Jackson, R.S.; Burse, V.W.;
     Hayes, C.; Straub, W.E.; Landrigan, P.J.; Liddle, J.A.
     Epidemic kepone poisoning in chemical workers. Am. J. Epid,
     1978, 107, 529-537.
3.   Christensen, H.E.; Fairchild, E.J. "Suspected carcinogens,"
     2nd ed., U.S.DHEW, NIOSH, Government Printing Office, Supt.
     of Doc., Washington, DC 20402.
4.   Crump, K.S.; Masterman, M.D. Review and evaluation of methods
     of determining risks from chronic low level carcinogenic
     insult in "Environmental contaminants in food," Congress of
     the United States, 1979, Libr. Congr. Cat. No. 79-600207.
     Supt. of Doc., U.S. Govt. Printing Office, Washington, DC
     20402. Stock No. 052-00300724-0.

5. Carter, C.D.; Kimbrough, R.D.; Liddle, J.A.; Cline, R.E.; Zack, M.M.,Jr. Barthel, W.F.; Koehler, R.E.; Phillips, P.E. Tetrachlorodibenzo-dioxin: An accidental poisoning episode in horse arenas. Science, 1975, 188, 738-740.
6. Dunphy, J.H.; Hail, A. Waste disposal: It's a dirty business. Chemical Week, March 1, 1978, pp. 25-29.
7. Brown, C.C. Mathematical aspects of dose response studies in carcinogenesis: The concept of thresholds. Oncology, 1976, 33, 62-65.
8. Dunphy, J.H.; Hall, A. Waste disposal: Settling on safer solution for chemicals. Chemical Week, March 8, 1978, pp. 28-32.
9. Guess, H.; Crump, K.; Peto, R. Uncertainty estimates for low-dose-rate extrapolations of animal carcinogenicity data. Cancer Research, 1977, 37, 3475-3483.
10. Hoel, D.G.; Gaylor, D.W.; Kirschstein, R.L.; Saffiotti, U.; Schneiderman, M.A.; Estimation of risks of irreversible delayed toxicity. J. Tox. & Env. Health, 1975, 1, 133-151.
11. Johnson, C.J. Toxic soluble waste disposal in a sanitary landfill site draining to an urban water supply. AJPH, 1977, 67, 468-469
12. Kreiss, K.; Zack, M.M.; Kimbrough, R.D.; Needham, L.L.; Smrek, A.L.; Jones, B.T. Cross-sectional study of a community with exceptional exposure to DDT. JAMA, 1981, 245, 1926-1930.
13. Schmahl, D. Combination effects in chemical carcinogenesis (experimental results). Oncology, 1976, 33, 73-76.
14. Wolff, A.H.; Oehme, F.W. Carcinogenic chemicals in food as an environmental issue. JAVMA, 1974, 164, 623-629.
15. "Chemical and Health Report, 1973." United States International Trade Commission Reports on Production and Sale of Synthetic Organic Chemicals, 1918-1976.

RECEIVED July 16, 1982.

Analysis and Risk Assessment: Key to Effective Handling of Hazardous Waste Sites

DONALD L. BAEDER

Occidental Petroleum Corporation, Los Angeles, CA 90024

The real key to effective management of hazardous waste disposal sites is:
- a responsible analytical base to accurately measure levels of potential or actual human exposure, which includes comparison with other relevant areas
- a comparative base of toxicological and epidemiological information
- an assessment of risk based on these two fact bases
- an independent check by qualified experts separated from political and emotional factors

Omitting one or more of these actions before releasing information directly to the public is an unconscionable risk that should never be taken. In fact, it could well introduce an even greater risk of irreparable psychological harm. Love Canal, which is discussed in detail, is the primary example where these actions were not taken by governmental agencies.

The real key to effective management of hazardous waste disposal sites is:
- a responsible analytical base to accurately measure levels of potential or actual human exposure, which includes comparison with other relevant areas
- a comparative base of toxicological and epidemiological information
- an assessment of risk based on these two fact bases
- an independent check by qualified experts separated from political and emotional factors

Omitting one or more of these actions before releasing information to the public regardless of what may be required to be reported to an agency, such as under Section 8(e) of the Toxic Substances Control Act, is an unconscionable risk that should never be taken. In fact, it could well introduce an

0097-6156/82/0204-0033$06.00/0

even greater risk of irreparable psychological harm. For the past two years, the need for an independent peer review as an integral part of Superfund has been foremost in my priority of things to accomplish. The evidentiary need for such a review is compelling. Unless it is done we cannot expect ever to effectively manage hazardous waste incidents. Unfortunately, my efforts to convince the committees writing Superfund failed.

Today I have the opportunity to lay before you the evidence that I feel is compelling to force this four-part approach to define human and environmental risk associated with hazardous waste sites -- and that is what I intend to do.

I know some of you will say "All well and good. When you have plenty of time. But what about the situations that require immediate action?" If you'll stay with me, I'll address this issue later, but I will say now that the necessary review can be done within the time frame required by the agency to make a decision.

I would like to flood you with examples of where proper actions were taken. Unfortunately, the record is dismal and so we will have to profit from our mistakes. I will use Love Canal as the primary example, not only because I am most familiar with it, but because it is clearly the best example. If nothing else, Love Canal has become almost a generic term.

Let's start with the events that led the State Commissioner of Health to call for a limited evacuation of pregnant women and children under two, which caused such an uproar among the residents that within a few days Governor Carey was forced to evacuate all the people in the first two rings of homes.

On October 3, 1976, the first confirmed reports issued that chemicals had seeped into the basements of some homes on the periphery of the Love Canal property. A government task force comprised of the City of Niagara Falls, the Niagara County Health Department and Hooker Chemical began to study the situation. The City acting as the lead commissioned a contractor to prepare a report which was received by the City in August, 1977.

In March 1978 the City commissioned an outside contractor to design a remedial program. Hooker participated in the study and offered to pay one third of the then-expected cost of remedial work which was estimated at $840,000.

On August 2, 1978, the New York State Health Department ordered the temporary closing of the 99th Street School and recommended the temporary evacuation of pregnant women and children under two living in the first two rings of homes around the Canal property during the completion of the remedial program. Approximately 20 families could have been affected.

On August 9, 1978, Governor Carey of New York visited the area and announced that all 236 families living on both sides of 97th and 99th Street (the streets bordering the Canal property) would be evacuated and their homes purchased.

The data base for this decision was two studies. One covered air sampling in the basements of several hundred houses by the state and eleven by EPA, the other a limited epidemiological study of adverse pregnancy outcomes in residents of the houses closest to the canal. These data can be used to support the need for the first two points mentioned above. That is:

- a responsible analytical base to accurately measure levels of potential or actual human exposure, which includes comparison with other relevant areas
- a comparative base of toxicological and epidemiological information

While an analytical data base was developed, the data was not evaluated in comparison with other typical locations. In addition, the epidemiology did not use an appropriate control group. Rather than use an equivalent local population not subject to the same exposure, the comparison was with the national averages. It is well known that there are marked variations in rates in small groups within a large population. For this reason, the control and test population should have similar size as well as background except for the thesis being tested.

Comparison of the highest value air samples from basements as tested by New York State and EPA and comparing the results with U.S. workplace standards (which are based on 8 hour exposure, not living at home) shows that the levels measured were orders of magnitude lower than the published workplace standards for the several common solvents detected.

COMPARISON OF HIGHEST VALUE AIR SAMPLE

CONCENTRATIONS INSIDE BASEMENTS WITH WORKPLACE STANDARDS
(parts per billion)

Compound	N.Y. State Highest Value	EPA Highest Value	Permissible U.S. Workplace Standards
Chloroform	5	3.0	10,000
Trichloroethylene	13	2.7	100,000
Tetrachloroethylene	170	7.0	100,000
Chlorobenzene	52	0.8	75,000
Chlorotoluene	1,300	40.0	50,000

These standards are set by an expert committee of the American Conference of Governmental Industrial Hygienists who

establish these values based on the available medical/technical
literature and their professional judgment.

Please note that these are highest values. The mean
levels are typically 1/10 to 1/1,000 of the highest values.
All samples shown were for basement air. Several ring 1 homes
were analyzed for chemicals on the first floor. Only in the
case of one home were measurable levels found on the first
floor. Chemical levels detected in basements of homes beyond
the first ring were sharply lower than for homes in ring 1.
The mean levels for ring 2 were 5 percent of the mean for ring
1.

In addition, data shows that outdoor air in the Love Canal
Area was comparable or better than in other major cities.

EPA CONTRACTOR DATA

COMPARISON OF ANALYSES OF OUTDOOR AIR
AT LOVE CANAL, RAHWAY AND PHOENIX
(parts per billion)

Compound	Love Canal	Rahway	Phoenix
Chloroform	7.3	5.2	0.1
Trichloroethylene	0.02	9.9	0.5
Tetrachloroethylene	0.3	3.5	1.0
Chlorobenzene	0.003	0.1	0.2
Chlorotoluene	0.09	*	*

* Not analyzed for

It is interesting to note that an article in the Buffalo
Courier Express dated September 23, 1978 headlined "Study Shows
'Canal' Air Beats L.A." tried to put exposure levels in perspective by comparing data for basement benzene and toluene
levels with atmospheric concentrations in Los Angeles, as
posted in the office of a Dr. Philip Taylor, then on-site
director for the state Health Department.

The benzene levels measured in basements of the first ring
of houses averaged only one fifth the outside atmosphere in Los
Angeles. Toluene levels in the first ring basements were 50%
higher. In the second ring, benzene was one tenth the Los
Angeles average and toluene one quarter.

As far as we know, this approach was not used again.

When one looks at the limited epidemiological study of
adverse pregnancy outcomes in residents of the houses closest
to the canal one finds the population considered was very small
since a total of only 236 families were involved. One segment
of the population showed a higher than average rate while the

other showed a lower than average rate, neither very significantly different. If combined, the rate was lower than the
national average. Panic rather than reason prevailed. The New
York Health Department justified its position by issuing a
booklet in September 1978 entitled, "Love Canal -- Public
Health Time Bomb," which described the Love Canal situation as
an "environmental nightmare" capable of causing "profound and
devastating effects" constituting a condition of "great and
imminent peril."

The inflammatory nature of this pamphlet and the concern
of the local residents was further exacerbated by a "study" by
Dr. Beverly Paigen, of Roswell Park Memorial Institute (RPMI)
in February 1979, urging the evacuation of families in light of
her study suggesting a high rate of birth defects.

The Paigen study was completed after the evacuation of the
first and second ring of homes. The Commissioner of Health did
not concur with her findings and stated that "we cannot say
with certainty that the higher rates found in each of the
categories are directly related to chemical exposure but the
data do suggest a small but significant increase in the risks
of miscarriages and birth defects. Although the magnitude of
the additional risk is indeed small, prudence dictates that we
take a more conservative posture to minimize even that small
additional risk." Based on this, he recommended temporary
relocation of children under the age of two and pregnant women
from the area beyond the second ring of homes -- another 700
families.

Note that Dr. Paigen did not do this "study" at Roswell
Park and it was outside her area of expertise. She is a
biologist and not a toxicologist. There was no peer review and
the "study" was taken at face value, further adding to the
panic.

It is clear that the key points mentioned at the outset
were not followed here since a comparative base of exposure
levels and epidemiological information was not used, nor was
there an independent check by experts separated from political
and emotional factors.

It was not until nine months later that the state made a
strong effort to develop a peer review. On June 4, 1980, New
York Governor Hugh L. Carey appointed a special blue ribbon
panel of distinguished physicians, chaired by Lewis Thomas,
M.D., Chancellor of Memorial Sloan-Kettering Cancer Center, to
"determine whether previous studies validly demonstrated acute,
chronic or long term health effects from exposure to chemical
wastes buried at Love Canal." On October 10, 1980 the Panel
issued its report concluding that it "...recognizes that there
was a reason for the State Health Department's initial
announcement of 'Public Health Time Bomb,' but not a good
enough reason. There ought to be a better mechanism for
convincing the Federal government that a certifiable disaster
area exists, in order to obtain Federal funds, than to arouse

such fears of imminent peril as swept through the Love Canal area in this case."

The Governor's blue ribbon panel, a year and a half later, long after the psychological damage had been done by the report, concluded that the Paigen report "...falls far short of the mark as an exercise in epidemiology. She [Dr. Paigen] believes fervently that her observations prove the existence of multiple disease states directly attributable to chemical pollution, but her data cannot be taken as scientific evidence for her conclusions. The study is based largely on anecdotal information provided by questionnaires submitted to a narrowly selected group of residents. There are no adequate control groups, the illnesses cited as caused by chemical pollution were not medically validated... The Panel finds the Paigen report literally impossible to interpret. It cannot be taken seriously as a piece of sound epidemiologic research, but it does have the impact of polemic."

A study with this many defects could easily have been initially reviewed and shortcomings noted in a few days.

The failure of the state to adequately review so-called studies was also shared by the federal government which added to the confusion when on May 17, 1980, the EPA held a press conference to release the results of a preliminary genetic study showing chromosome damage of 11 of the 36 Love Canal residents tested. The EPA recognized that there was no control group and that "prudence must be exerted in the interpretation of such results." The report also stated "we strongly recommend the cytogenetic analysis of a larger population of Love Canal residents along with a number of unexposed individuals (controls) before significance can be placed upon the results." Hooker itself called for an immediate followup on the EPA study.

The federal government was able to obtain immediate peer review and on May 21, 1980 a special panel convened by the U.S. Department of Health and Human Services in essence blasted the EPA genetic report. The panel stated that "It provides inadequate basis for any scientific or medical inferences from the data (even of a tentative or preliminary nature) concerning exposure to mutagenic substances because of residence in the Love Canal area." The panel also indicated that "we do not believe that on the basis of this report it should be concluded that the chemical exposure of Love Canal may be responsible for much of the apparent increase nor can we concur with the report's implication that a cytogenic observation suggests that the residents are an increased risk of neoplastic disease, of having spontaneous abortions and of having children with birth defects based on the evidence presented."

The EPA study was inadequate, but again the release three days earlier of an inadequate report had created other major problems.

Note that the EPA study was commissioned by government attorneys building a legal action, and selected subjects for the study not from a normal cross-section of the public at risk. "The EPA study, however, was not well designed. It was not even meant to be scientific, according to Stephen Gage, assistant administrator for research and development at EPA. 'This [the study] was a small fishing expedition. The Justice Department asked us to undertake it in connection with our suit against Hooker,' he says." 208 Science, June 13, 1980, p. 1239. The EPA "chromosome" debacle so inflamed the local residents that they kept two EPA officials hostage for several hours. Lois Gibbs, the president of the Homeowner's Association, indicated that "this action was a 'direct result' of the study being released with nobody to tell us what it meant or what they were going to do about it! If the EPA officials had been let out the front door, 'they would have been torn apart.'" Love Canal Residents Under Stress, Vol. 208, Science, June 13, 1980, at p. 1242.

A "political" decision issued on the same day as the peer review blasted the EPA study. After the EPA officials had been held hostage, President Carter declared a state of emergency in the Love Canal area, paving the way for evacuation of up to 710 families.

Governor Carey concluded in a newspaper interview on May 31, 1980 that "a costly relocation of more than 700 Love Canal homeowners is medically unnecessary but had to be carried out to assuage the panic caused by the EPA report." The Director of the State's Office of Public Health said that based on studies going on since 1978 the frequency of birth problems of the rest of the Love Canal (other than the first 230 relocated families) was comparable to that of Niagara County."

Again we see a significant community effect resulting from a failure to properly manage the assessment of risk. Information was released based on an inappropriate analytical base without adequate epidemiological information or independent review.

On June 24, 1980, the New York Department of Health Commissioner finally issued the Department's epidemiological studies on Adverse Pregnancy Outcomes in the Love Canal Area, concluding that "we have not yet been able to correlate the geographic distribution of adverse pregnancy outcomes with chemical evidence of exposure. At present, there was no direct evidence of cause-effect relationships with chemicals from the Canal." The Commission also concluded that "comprehensive studies of three households with unusually adverse reproductive histories did not produce evidence of unusual risk of chemical exposure."

The studies also concluded that:
"Blood testing, which was designed to screen for liver and kidney abnormalities, leukemia and blood diseases, showed no patterns of excess abnormality."

"None had clinical evidence of liver disease."

"Computer analyses of the 22-page health questionnaire, which elicited information on some 150 different diseases or symptoms, produced no evidence of unusual patterns of illness or other disorders. Cancer incidence was within normal limits for this population."

Attempts to determine the basis for EPA decisions to relocate this massive group was unavailing. Congressman John J. LaFalce (Democrat, 36th District, New York) representing the Love Canal area, asked for the data base justifying the decision and was told it was part of material being used in pending litigation. In a letter to President Carter dated June 20, 1980, Congressman LaFalce stated "I find no justification whatsoever for making broad policy judgments -- such as the one here to recommend temporary relocation of 2,000 people -- based on information developed in conjunction with litigation, and then refusing to make that information available to the people whose lives are devastated by it, and to public officials who must make public policy judgments hopefully based on fact rather than surmise."

Congressman LaFalce also summarized clearly the policy issues involved with Love Canal in his letter to President Carter of June 20, 1980:

> The Love Canal is a unique situation.
> No health authorities at any level of
> government have ever conducted
> comprehensive health tests in accept-
> able scientific manner, despite my
> repeated call for them. Some resi-
> dents of the neighborhood believe
> that they suffer from very serious
> health problems and that the cause is
> their exposure to toxic chemicals.
> Others believe that there is no more
> risk in the area than in many other
> urban environments. Still others who
> do not know what to think and have
> been suffering terrible mental
> anguish for the past three years.
> The one thing these three groups do
> have in common, however, is a lack of
> information thanks in part to the
> attitude of the United States Depart-
> ment of Justice. A whole neighbor-
> hood is stigmatized to the point that
> the property there is literally
> valueless. The people who live there
> are frightened beyond belief and an
> arrogant official in the Justice
> Department says they cannot even be
> told why this has been done.

The data base was not forthcoming to Congressman LaFalce or to anyone else. Some cynics might argue that perhaps this campaign was linked to then pending Administration efforts to pass Superfund. Whatever the basis, as the New York blue ribbon panel concluded in October, 1980, "much of the anxiety caused for the Love Canal residents might well have been averted if a <u>single</u> Federal-State group had evolved early in the history of the situation and that public pronouncements were made only by this group and limited to the exactitudes permitted by the current state of scientific knowledge. The scientific evidence, incomplete though it is, reveals no state of population damage justifying the term 'imminent peril' and 'profound and devastating effects.'"

After a lengthy review of how the Love Canal situation was handled by the various governmental agencies, the Panel members concluded that:

> "...The record indicates just as clearly that an articulated and coordinated Federal and State approach was not achieved and has not been achieved to date.... There did not emerge at any point in the past two years anything like a master plan for assembling the kinds of information required for analysing and comprehending the problem...because of this ambiguity, the people most directly affected by the conditions at Love Canal have been subjected to more than two years of the most intense anxiety and fear. In the absence of clean-cut, authoritative answers, many of the residents have come to believe that their health is in fact irreversibly damaged, that they are at future risk of cancer, congenital malformations in their offspring, and an increased incidence of miscarriages and abortions."

As for the studies the Panel report indicated "...the publication of studies on health effects and subsequent criticisms of these studies in the media have created more uncertainty than understanding on potential health problems for both the public and government officials...The inadequate coordination of study designs and procedures to insure meaningful findings concerning health effects has exacerbated the problems faced by decision makers in responding to this situation."

Finally, as to whether there were any acute or chronic health problems from exposure to the chemicals in the Love Canal, the Report concluded:

It is clear enough from the
available data that no acute cases of
intoxication by chemical pollutants
have been observed within any part of
the Love Canal community, 'wet' or
'dry.' That is, no clusters of cases
of acute liver disease, or kidney
disease, or pulmonary manifestations,
or hemolytic anemia or agranulo-
cytosis, and certainly no peripheral
or central nervous system syndromes.
Whatever else may be going on, there
has not been a sufficient concentration of toxic material to produce
overt illness attributable to poisoning.
This was clear enough from the
outset.
. . . .
As a result of this review, the
Panel has concluded that there has
been no demonstration of acute health
effects linked to exposure to hazardous wastes at the Love Canal site.
The Panel has also concluded that
chronic effects of hazardous waste
exposure at Love Canal have neither
been established or ruled out yet, in
a scientifically rigorous manner.
The studies conducted in the past two
years have been inconclusive in
demonstrating long term health
effects due to hazardous waste
exposure.

As this talk is written, we are still waiting for a final
EPA report on the health impacts of Love Canal.

After review of the fact chronology in Love Canal it is
easy to see that the factors for effective management of
hazardous waste disposal sites were simply not utilized or
available in Love Canal. These factors again are:

- a responsible analytical base to accurately measure
  levels of potential or actual human exposure which
  includes comparison with other relevant areas
- a comparative base of toxicological and epidemiological information
- an assessment of risk based on these two fact bases
- an independent check by qualified experts separated
  from political and emotional factors

Love Canal is the best or worst example of the failure to
use common sense in dealing with a potential alleged risk to
human health. But another example, particularly of psycho-

logical impact, is detailed in an article entitled <u>The Dump</u>
<u>That</u> <u>Wasn't</u> <u>There</u>," 215 <u>Science</u>, February 5, 1982, at p. 645,
cited as an example of the perils of relying on residents
reporting an increased incidence of health effects. The facts
as reported in <u>Science</u> are as follows: A resident in a blue
collar residential neighborhood in Memphis called the local
health department in 1976 to complain of rashes and other minor
illnesses, alleging that she and her family were being poisoned
by toxic chemicals in the environment. Investigation of her
home and yard disclosed only trace concentrations of chlordane
and other pesticides that are used in Memphis to control
termites and mosquitos. The same woman called again in 1977
and 1978 with the same complaints and same results.

By the summer of 1979, other residents in the neighborhood
had begun to complain of increased incidence of rashes, headaches, urinary problems, heart disease and cancer, complicated
by the allegations of one former health department employee
that he "knew" the location of a chemical waste dump in the
area. By the following April, the Director of the health
department indicated that there was a "highly charged atmosphere, virtually identical to that at Love Canal."

The citizens "panicked" and there were picketing and
emotional meetings. There was talk of evacuating the area and
local political activists pressed for immediate action in the
area. There were even Congressional hearings. Extensive
investigation by EPA found "absolutely nothing higher than the
background levels of pesticides." A respected professor at
Johns Hopkins University went through old records and conducted
other studies with no indication of the source.

The U.S. Center for Disease Control conducted a door-to-
door survey of some 300 homes asking about effects and examined
physicians' records. The survey "did not produce any evidence
of toxic illness or severe health effects." However, the
survey did show a slight increased incidence of headaches and
skin rashes, but the conclusion was that it was "unlikely they
were chemically caused." One man found to have heavy metal
poisoning worked in a foundry and one of the children of the
woman making the original complaint was found to have hives.
It took 14 months for the Center for Disease Control to publish
a study and by that time, according to the head of the local
health department, "nobody believed any agency." About the
same time the local authorities did become aware of a local
dump site, but it was some 3 to 4 miles away from the neighborhood and separated by several physical barriers so "that there
is no chance of any chemicals from it having reached Frayzer
[the local neighborhood]."

There are still a small number of people in the neighborhood who believe they have unusual health problems, but by now
the uproar has died down. The article concludes with the
comment that "the specter of this ghost dump will continue to

haunt epidemiologists confronted with other self-reported
increases in illness."

How can we prevent the crying of "wolf" and the unnecessary psychological distress caused by government actions at
Love Canal and the "phantom dump?" These problems are real and
are described in detail in Love Canal Residents Under Stress,
208 Science, June 13, 1980 at p. 1242, which graphically
discusses the psychological impacts of the uncertainty created
by the muddled governmental response.

Perhaps the best way is for federal and state governmental
agencies to have a peer panel available on call when there is
an incident, to review data before it is released. Having the
panel available will save valuable time in the event that there
is an imminent danger. Such panels can act quickly, as, for
example, in the review of the EPA chromosome study in a three-
day period. Unless the federal and state governments have a
mandate in regulations and law to have such peer review panels
available, we cannot ever expect to effectively manage hazardous waste incidents. The decisions in this area are too
important to allow an agency under political and emotional
stress to make this decision alone. Confidence in government
is significantly eroded when the data are not available or lack
a scientific basis. Confidence in industry becomes non-
existent when such action is taken. Love Canal has become a
generic term for health effects on individuals and irresponsi-
bility of corporations, although ironically the Task Force on
RCRA of the American Institute of Chemical Engineers concluded
that "the design of the Love Canal site was well within the
standards of RCRA," long before that statute was promulgated.

The facts indicate that the characterizations are unfair,
but the public remembers the concerns raised by the agencies
and the press and not the scientific base. Unfair stigmatiza-
tion of companies is wrong, but traumatizations of whole local
populations without a proper basis is reprehensible.

RECEIVED June 16, 1982.

An Industry's Guidelines for Risk Assessment

RICHARD H. DREITH

Shell Oil Company, Houston, TX 77210

In the absence of a consensus on absolute limits
of risk, it is useful to have a goal-oriented risk
assessment framework for decision making and
response resource allocation. A four-step process
is in use at Shell to evaluate risks and define
appropriate responses. The steps are: (1) Hazard
Identification, (2) Hazard Evaluation, (3) Risk
Evaluation, and (4) Risk Response. The first three
steps amount to making a "risk assessment," and the
fourth step adds a response definition. A risk
classification system setting high, low and insignificant categories for risk reflecting unacceptable,
variously acceptable and acceptable regions respec-
tively are defined, and companion levels of response
action are presented. An overview of a site evaluation experience in California is outlined.

Risk assessment is a difficult problem area being addressed
by industry and regulatory agencies today. I have found in my
professional experience in the oil and chemical industry, that
when faced with solving difficult problems, solutions generally
resulted by taking logical and disciplined approaches to resolve
uncertainties. We at Shell have developed and are using guidelines for risk assessment which are allowing us to analyze risks
and identify reasonable responses. The use of these systematic
procedures have improved our confidence in the decisions we must
make. In responding to risk situations, our confidence is built
on knowing that the responses are directionally correct and are
achieving our overall goal to reduce risk.

Of late, I have been disappointed by the consensus opinion
which appears to be growing regarding dealing with risk assessments. From information in the media and discussions with con-
temporaries and agency representatives, I detect an opinion that
suggests society is not ready to deal systematically with

0097-6156/82/0204-0045$06.00/0

assessing risk. In my view, those who support this claim are as
remiss at that extreme as those who would now place total faith in
seemingly highly precise methodologies for extrapolations and risk
assessments. Additionally, the opinion follows that we are not
ready to translate current risk assessment understanding into
regulatory programs being implemented.

I don't think we at Shell agree with either opinion, or are a
part of any consensus embracing these opinions. We will be disappointed, I'm sure, if the implementation of either the Resource
Conservation and Recovery Act (RCRA) or the Comprehensive Environmental Response Compensation and Liability Act (CERCLA) programs
proceed without there being incorporated at least some regulatory
allowances for assessing actual risks. The risks posed by various
hazardous waste handling activities must be assessed when judgments on their acceptability are made, such as during the permitting process. I am suggesting that the ability to assess actual
risks exists, and it should be considered when (1) setting regulatory requirements, (2) when establishing facility performance
standards, or (3) when defining remedial activities.

We fully recognize that many areas of uncertainty are confronted when conducting risk assessments. These include the lack
of generally accepted specific assessment techniques. Furthermore, even when risk is assessed there is a lack of a consensus on
the absolute limits of acceptable risks. We do feel, however,
that there are systematic approaches that can and are being used
and, at the minimum, they are directionally correct and can serve
as a conceptual framework for regulatory program development.

Shell Risk Assessment Procedure

I will outline very briefly for you the guidelines, process
and procedures used at Shell for risk assessment and risk response (1), (2). I will focus my comments on how we are beginning
to deal, for example, with decisions concerning the acceptability
of operating existing hazardous waste disposal facilities or when
considering new waste handling facilities.

The step-wise evaluation and decision process can also be
used in assessing the environmental risks posed by abandoned
hazardous waste disposal sites. I will also briefly describe our
experience with a site in a southern California community.

Our Risk Assessment Procedure has grown out of a continuing
need to deal with balancing all the aspects of health and
environmental risks with all other pertinent elements in our
oil and chemical business activities. The decisions we must make
to deal appropriately with risk are necessarily an important
consideration in pursuing our overall business plan.

In many cases, we must take action and make decisions prior
to regulations being issued which deal with risk in specific
areas. By necessity then, we need and use a system that reflects
our corporate environmental goals and also allows guiding the

allocation of the limited resources available to reach those
goals.

In a sense, we can say we have developed a system to assess
environmental risk situations and are responding to those assessments with decisions reached in a deliberate and disciplined
manner. The confidence we have in the system is built on the
premise that we know it is directionally correct with its overall
objective being to reduce risks. Moreover, the response decisions called for in the process result from evaluating all the
available scientific information along with all the pertinent
values and judgments that can be brought to bear.

Overall, the major premise for our approach is a
directional orientation toward risk reduction. The evaluation
procedure used follows a four-step process which considers first
Hazard Identification; second, Hazard Evaluation; third, Risk
Evaluation; and fourth, Risk Response. To avoid any misunder-
standing of terms, the combined activities of the first three
steps can be considered as what is commonly referred to as making
a "Risk Assessment." The fourth step, Risk Response, necessarily
must follow when the process is used to make practical decisions.

I will now give an overview discussion identifying the
general information needs and the specific decisions called for at
each step. I will use as the example case, evaluating a hazardous
waste disposal site. Also, I will discuss the application of the
process to an actual site being evaluated in California.

Hazard Identification

As is typical in regulatory matters, understanding the
definitions of the terms used is very important. First, in
the broadest sense, hazard means the potential to do harm, or in
other words, the potential to cause an adverse effect.

In a waste disposal site evaluation, the hazard identification step will involve collecting and validating all recorded and
other information on the nature and properties of the wastes
actually in a site. The purpose is to determine whether a hazardous situation actually exists. Stated simply, is there a potential for harm to health and the environment when considering the
materials present?

The type of data gathered will include all pertinent information on (1) the inventory of the wastes disposed of at the site,
(2) the composition of those wastes, (3) the physical and chemical
properties such as persistence of compounds and their solubilities,
(4) the biological properties such as toxicity, and (5) the potential interaction of wastes and degradation products.

The decision called for at this point in the process centers
on confirming that, from the qualitative data gathered, a potential for an adverse effect has or has not been identified. If no
hazard exists, further action is not warranted. If the potential
for an adverse effect is evident, the process requires proceeding
to the next step - Hazard Evaluation.

Hazard Evaluation

The hazard evaluation step involves determining the extent
or scope of the potential adverse effects from the hazards
identified in step one. The evaluation effort now progresses to
include analyzing and describing both <u>qualitatively</u> and <u>quantita-
tively</u> the characteristics which might cause the wastes to be
hazardous and to providing the means for determining in step
three, the risks involved under the disposal circumstances specific to the site being studied. These means may include gathering
pertinent information from laboratory and other studies on the
toxicological properties of the array of constituents in the site.
Also pertinent would be the range of responses to those constituents under the various stated, not actual, levels and types of
exposures along with reviewing inter-species translation relationships and the means to extrapolate these to actual conditions.

With the focus on the actual quantities of wastes identified
in the site inventory, appropriate field sampling and analysis
must be undertaken to verify that the identified hazards actually
exist. If a hazard has been identified and the potential for
causing adverse effects confirmed, after considering both the
qualitative and quantitative dimensions of the situation, the
process requires proceeding to the next step - Risk Evaluation.

Risk Evaluation

Risk evaluation in the assessment process focuses on <u>weighing
the health and environmental threats posed by an identified</u> haz-
<u>ard</u>. First, the possibility and probability that a person will
actually experience an adverse effect as a result of the existence
of the environmental hazard and exposure to it must be weighed;
and, secondly, the number of persons who might be exposed must be
considered.

In the stepwise process, the risk evaluation combines the
results of the second step, hazard evaluation, with any information on actual exposure possibilities, including evaluating exposure sources, levels, frequencies, types and routes. The assessment effort involves interpreting the field verified data from the
perspective of determining what the <u>actual</u> risk level to humans
and the environment is in the <u>real world circumstances</u> posed by
the activity being evaluated.

In the case of a hazardous waste disposal site, the geohydrologic setting characteristics must now be carefully considered to assess the existence of subsurface exposure routes.
Also, the potential for exposure from air emissions and surface
runoff must be quantified where possible. Overall, the significant factors would include: (1) evaluating the volatility of the
wastes which could lead to short- and long-term exposure to air

emissions, (2) considering the long-term solubility of the wastes
in waters from either rainfall or surface runoff, (3) considering
the continuing potential for surface water contact, (4) reviewing
the rainfall rates and rainfall flow paths as to their ability to
support their contribution to continued exposure, (5) identifying
the groundwater contamination potential and migration paths, (6)
identifying groundwater uses, and (7) appraising population exposure by considering proximity and number.

If the risk evaluation step shows that a significant potential for adverse effects exists, the stepwise process calls for
developing an appropriate response. Determining the appropriate-
ness of a response implies having a basis to vary the response.
Providing that basis is the end product of the risk evaluation
step. It involves classifying the risk as being high, low or
insignificant so that the appropriate response can be developed.

Before discussing how risks are classified, I would like to
emphasize that the hazard identification, hazard evaluation, and
risk evaluation steps are purposely kept free, as much as possible,
of value judgments. Credible risk evaluations necessarily require
using the best scientific input, theory, acumen and judgments. In
the fourth step, value judgments come into play to assist in
developing risk responses reflecting those societal judgments that
bear on the situation being evaluated.

To classify risks, a system or framework for classification
must be agreed upon. In a perfect world, a societal consensus on
levels of "risk acceptability" would be at hand and be straightforward in how it would be applied in each situation being
evaluated. However, in the real world there is no consensus, and
we can only evaluate and then define the ranges of risks posed in
actual situations and then judge them for their acceptability to
society.

Within Shell, when taking actions beyond those required by
law, we use the concept of risk regions or risk classifications.
The objective of our classification scheme is twofold; first, to
provide guidance for the allocation of resources such that the
most serious risks are handled first; and second, to identify
those situations where allocating resources is not warranted since
the risk is judged to be insignificant.

Our system, therefore, distinguishes three classifications of
risk: high, low, and insignificant. One might argue that a
medium risk region exists. At this time, considering a medium
risk region would only add an unrealistic aura of precision to the
system. We recognized at the outset that the tools available to
us in our classification scheme are blunt.

Definition of the boundaries between the risk regions can be
precise by choice; however, there are no established criteria for
boundary setting and, at this time, only guidelines for boundary
selection can be offered (1). The range of choices might include
(1) reasonable performance goals for best technology control
equipment when actual exposure levels from existing facilities is

assessed, (2) the feasibility of controlling exposures, and (3) a
desirable industry or activity performance goal.

While the framework for risk regions is precise, the uncertainties emerge when real world imprecise evaluations of risk -
recognizing the bluntness of our tools - are measured by the
precise yardsticks. Nevertheless, when considering risk classifications from a societal viewpoint, one can have confidence and
logically classify high risk regions as being regions of "unacceptable risk." The low risk region might similarly be labeled as
"variously acceptable" with the third region referred to as
"insignificant risk." Consistent with that classification scheme,
one could define a parallel series of appropriate responses.

Risk Response

Risk response involves additional difficulties. Defining
responses includes making comparisons with other examples of
risks society takes, including their view on the seriousness
of the risk and the perceived benefits from taking that risk.
From the process of combining the scientific information available
with the pertinent value judgments, decisions can be made as to
what actions are required and justified as appropriate response to
the risk posed.

When a risk assessment shows a "high risk" situation, this
would be judged unacceptable and require a response to abate the
risk and move it downward toward the "low risk" or "variously
acceptable" region.

Similarly, an "insignificant risk" classification is judged
as defining those situations which do not justify or warrant the
allocation of resources for further risk reduction. I must emphasize at this point that the "insignificant" classification is not
equivalent to an absolute "zero risk."

The "low risk" or "variously acceptable" region under this
labeling scheme would then, by difference between high and insignificant, identify those situations where balancing decisions
which consider risks, benefits and justification must be made for
prioritizing the application of critical resources.

The exercise of balancing risks and benefits in the "low risk"
region is clearly an area where regulatory guidance should emphasize the importance of relating risk reduction to costs, benefits
and alternative risks. This "low risk" or "variously acceptable"
region is the area where costs to reduce risk should be justified
and alternatives considered. For just as zero risk is not
generally attainable, very little real progress toward our goal
would be reached if we attempted to shift all "variously acceptable" risks downward.

Regardless of the absence of a governmental policy establishing absolute risk limits or boundaries of acceptability, risk
response decisions must be made in many industrial situations. We
have found that coupling disciplined risk assessment procedures

with the risk region classification techniques yields appropriate
response actions, and the process has proven to be a <u>practical</u>
tool to deal with these inherently imprecise situations.

Using the risk assessment and classification approach in
evaluating waste disposal site situations should likewise allow
making practical decisions by coupling risk levels with appropriate remedial responses. The stepwise decision process should
also be useful in evaluating the scope of risks posed by abandoned
disposal sites. If the risk assessment process leads to the conclusion that a disposal site is, indeed, posing a <u>significant</u>
threat to health or the environment, appropriate levels and types
of responses can be defined. By considering all the quantifiable
factors, the identified uncertainties, and the geohydrologic predictions and projections specific to each site, supportable
practical response decisions can be made. If a "high risk" classification is supportable, the response action to reduce that
particular risk is progressed without comparison to other risks.
If an "insignificant risk" classification is given, no action is
required solely to reduce risk to health or the environment.

A circumstance worth mentioning at this point is the response
appropriate to deal with <u>nuisance</u> <u>and</u> <u>esthetic</u> issues such as odor
or noise when no hazard can be identified. A risk classification
of "insignificant," with respect to health or the environment,
would likely be selected and no response justified in the case of
a disposal site evaluation where only odor is the problem. In
these instances, appropriate responses should be developed using
a process which reflects the economic, social and political
impacts of the problem rather than the assessment process being
described which has been developed to deal with hazards to health.

In disposal sites with identified hazards, the more difficult
and involved decisions will have to be faced when the circumstances result in selecting the "low risk" classification. Since
total cost/benefit considerations appropriately come into play --
for here we hope to achieve the greatest reduction in risks for
each dollar spent -- the response actions can vary accordingly.
While risk reduction is the overall objective, reducing the selected risk level into the "insignificant" classification is not
automatically the optimum solution since it may require inordinate
and excessive resource utilization.

California Site Study

The evaluation of an actual disposal site in southern California has been following a pattern of events that, in ways, fits
the process I have been discussing. I would like to share with
you the highlights of this evaluation.

Shell is one of a number of participants in an effort to
define and resolve the problems posed by a disposal site in an
established community. The site was permitted during the 1940's,
and used for disposal of acidic materials from petroleum refin-
eries in the area. It is in a developed residential area.

As might be expected, when the site was identified and some of the circumstances surrounding its past use were exposed in the media, the reaction of many was to expect an immediate clean-up. Relating this reaction to the step-wise procedure I have been discussing, some jumped ahead to step four - Risk Response - and called for an immediate excavation without having considered steps one, two or three. Calling for immediate clean-up is equivalent to classifying the situation as one posing a high risk. Doing so can only be conjecture when only limited and qualitative data is at hand. This reaction was, of course, premature and disregarded completely such functions as identifying and evaluating the risk associated with excavation.

A better approach to resolve community concern is to move forward and follow the systematic procedure to first identify the hazard, evaluate its scope, assess the risks posed, and then develop an appropriate response.

At the California site, an early preliminary evaluation by federal, state, and local agencies was made to identify the nature of the materials in the site and the potential for an adverse effect. Then, with industry participation, a plan was followed to develop and then implement a detailed and comprehensive site evaluation.

The data from the evaluation effort will allow making a supportable decision on the hazards posed by the site. Also from the data gathered during the study, predictions and projections can be made as to the scope of both the current and future risks posed by the site. In addition, this process will provide information necessary to understand the risks associated with the specific remedial measures available and under consideration.

From consultations with the agencies involved, the community, and with the industrial participants, the risks posed by the site should be classifiable and appropriate response and remedial measures agreed upon. We have confidence that this scenario will be followed to reach a satisfactory resolution of the problems associated with this site.

Summary

I trust that you will agree that the use of a step-wise process for risk assessment and response development and the companion concept of using risk regions or risk classification, offers a framework for practical decision making.

Making the effort at this time to define boundaries between high, low and insignificant risks in risk assessment situations, is not preempting any ultimate societal judgement. Eventually governmental policy will likely define the risk boundary limits. In the interim, using systems and approaches that are directionally oriented to reduce risk can serve as a base for establishing the logical framework that facilitates practical decision making.

It would appear to be worthwhile to support efforts to develop risk assessment approaches along the lines I have discussed rather than join a consensus saying that it cannot be done, or in supporting selecting mathematical models which <u>arbitrarily</u> define hazards and risks.

We solicit your consideration in developing a broader consensus which supports the premise that a deliberate, disciplined, stepwise approach to making risk assessments is available for practical application in disposal site evaluations.

Literature Cited

1. P. F. Deisler, "Dealing with Industrial Health Risks: A Stepwise, Goal-oriented Concept," Chapter 15, AAAS Special Symposium #65, "<u>Risk in the Technological Society</u>," Ed. by C. Hohenemser and J. Kaspreson, Westview Press, Boulder, CO, 1982.
2. P. F. Deisler, "A Goal-oriented Approach to Reducing Industrially-Related Carcinogenic Risks," <u>Drug Metabolism Reviews</u> - Vol. 13, No. 5, Sept. 1982 - In Press.

RECEIVED June 16, 1982.

Methodology for Assessing Uncontrolled Site Problems at the County Level

RICHARD A. YOUNG

State University College, Department of Geological Sciences, Geneseo, NY 14454

ANN B. NELSON and LOUISE A. HARTSHORN [1]

Monroe County Environmental Management Council, Rochester, NY 14614

A comprehensive method to inventory inactive waste disposal sites should integrate all available historic, engineering, geologic, land use, water supply, and public agency or private company records in order to develop a complete and accurate site profile. Detailed information on site contents may not be available but areas of potential impact can be evaluated. Where people or drinking water supplies are affected, further investigation is indicated.

Historic aerial photographs provide the accuracy and documentation required to compile a precise record of site boundaries, points of access, and adjacent land use. Engineering borings for construction projects in the vicinity of suspected sites are integrated with geologic information to construct reasonable hydrogeologic models to evaluate potential leachate impact on water wells or nearby inhabitants.

Sites are systematically ranked using matrices with appropriate variables, such as geology, land use, or distance to water wells. This kind of an evaluation is a necessary step in the prioritization of sites where little is known about contents and where numbers of sites preclude a comprehensive drilling and testing program.

The Monroe County experience indicates that a well-designed study provides a conservative estimate of the number of large dump sites that require further consideration. This study provided a comprehensive, fifty-year inventory of all potentially significant sites in a large urban area (Rochester, New York).

[1] Current address: 65 Broad Street, Room 203, Rochester, NY 14614

0097-6156/82/0204-0055$06.00/0

The methodology described in this paper to locate and classify abandoned dump sites was developed in Monroe County, New York in response to a 1978 county legislature request to locate potential "Love Canal" type problems and a 1979 New York State law requiring counties to identify abandoned sites. The system is designed to evaluate and prioritize a large number of sites for more detailed investigation. While not all points will be applicable in every part of the country, the general approach will allow government agencies to focus limited resources on those sites that pose the greatest potential impact on human health.

This paper is a brief description of a methodology that will be fully described in a final report to EPA to be completed in 1982.

Rationale And Approach

The highest priority for the Monroe County study was to locate sites where potentially hazardous substances in high concentrations might be present in areas occupied by large numbers of people. There are many other situations where hazardous materials might have a potential long-term effect on humans or the environment that were excluded from the highest priority category in the design of the study, i.e., the slow, long-term release of toxic substances into streams and lakes through groundwater might be of great concern in other circumstances. However, the immediate effect of these low concentrations on areas where large numbers of people live or work might be negligible. Furthermore, detection of low concentrations of hazardous substances in water bodies used for drinking water or recreation is an area of current concern for health or water quality agencies and should be handled under existing programs. Information bearing on such long-term problems is not ignored in the present study but would be referred to the appropriate agency when encountered.

On the other hand, public or private well water supplies in close proximity to existing or abandoned landfills are placed in the highest priority category because the potential for a more direct contamination pathway is obvious.

An integrated study such as the one undertaken in Monroe County documents a great deal of historic information that could eventually be used for many purposes. It provides comprehensive baseline data that might be unrecoverable in future years. Planning activities and zoning requirements are obvious areas of impact. The information, once collected, could be used by any agency whose efforts or priorities might overlap but differ from the specific goals of the present study.

The key to the successful implementation of this type of program is the careful organization and integration by qualified individuals of all the critical resources that might be overlooked or ineffectively utilized in a less comprehensive effort with more limited time constraints. Involvement of personnel from all affected or peripheral agencies is strongly recommended.

Historic aerial photographic analysis is the best way to confirm actual or potential site boundaries. This, in turn, allows personnel and resources to be focused on those areas where physical documentation of sites is reasonably accurate.

The geologic analysis aids in both the search for and prioritization of potential sites, but the data must be integrated with the information developed by record searches, interviews and the careful development of a comprehensive site activity profile.

Finally, this effort is not meant to replace those technical studies which have dealt with the analysis and prioritization of known hazardous waste sites where a great deal of specific information is available on site history, contents, and hydrogeology. The present study should reduce the need for expensive drilling and testing programs and provide a means for their rational implementation.

Site Identification

In conducting a general survey of abandoned dump sites, there are many sources of information which must be used in combination to identify and/or verify site locations and to characterize site activity. The Monroe County methodology uses historic aerial photographs in stereoscopic pairs as the primary resource to develop an information base, but no single source should be depended upon entirely. Aerial photographs are the most reliable resource for identifying site boundaries, periods of operation, site drainage, and pre-existing uses or features such as sand and gravel operations, borrow pits, surface ponding, and access routes. However, it is not always possible to distinquish dumping from "clean" fill or construction activity on photographs. The information must be supplemented by other sources. Reports, maps, records, interviews and information from the general public all provide information that is useful in constructing a site activity profile. These sources, used without air photo information, may be insufficient to locate sites adequately and generally do not provide accurate information on site boundaries. It is the combination of all available resources that is the key to accurate site characterization.

Data Sources. The useful sources of data include:
. Aerial photographs in stereoscopic pairs for selected years
. Records from local and state health department and conservation agencies
. Environmental atlases
. Government publications on hazardous waste sites

. County soils maps
. Historic resources (newspaper clippings, articles, city
  directories, fire insurance maps, plat books, industrial
  surveys)
. Interviews with public officials, agency employees
. Public call-in campaign.

   General Survey. The general survey is conducted for one
USGS quadrangle sheet at a time. Historic aerial photographs
are interpreted for selected years (generally ten year intervals)
to identify abandoned dumps. A record of interpretations is made
on mylar sheets attached to one photo in each stereo pair. These
overlay sheets are retained as important permanent documentation
for the study.
   Since it is not always possible to distinguish between construction, filling, and dumping on the photographs, sites are
initially given one of six site activity designations (Table I).
Sites in the "Possible Dump" and "Unspecified Fill" categories
will need supplemental information to determine whether dumping
actually occurred. The mixture of types that occurred for one
quadrangle in Monroe County is shown in Figure 1.
   After the photos are interpreted for each selected year,
the information from all years must be combined for each site.
A written record of site activity is compiled and recorded. A
maximum site boundary map is prepared using large scale (1"=200')
maps. Orthophoto maps are best for this purpose if available.
Lastly, a general survey map at the scale of 1:24000 is completed
for each quadrangle by transferring boundaries from the larger
scale maps (Figure 1).
   Sites are classified for current land use from the most
recent set of aerial photography. One of six land use categories
is applied. These will eventually need to be verified by field
inspection.
. 24-hour occupancy on or within 100 feet of the site (residential use, hospitals, nursing homes)
. Part-time occupancy on or within 100 feet of the site (commercial establishments, industry)
. 24-hour occupancy within 1,000 feet of site
. Part-time occupancy within 1,000 feet of site
. 24-hour occupancy within 1,000-2,500 feet of site
. Part-time occupancy within 1,000-2,500 feet of site

Geologic Analysis

   A general geologic analysis of the entire region under study
is essential to the development of a rational approach to site
prioritization. Such a broad-based analysis also serves as the
frame of reference for site specific studies during the later
phases of the project.
   The important components of a general geologic analysis are

TABLE I.   Site Activity Categories

D-Identifiable:
    Sites where information on dumping activity is known from
public records, interviews with government or industry officials,
the public call-in campaign, industrial surveys, or where dumping
activity is clearly evident on aerial photographs.

P-Possible:
    Sites where filling activity is evident but there has been
no confirmation as to whether or not dumping has occurred. However, based on the location of the site and peripheral land use,
it would appear that dumping could have occurred. Sites located
adjacent to industrial or commercial activities, maintenance
areas, large construction sites and public facilities such as
sewage treatment plants and incinerators should be evaluated as
possible dumps.

U-Unspecified:
    Sites that are apparent either as recent surface distur-
bances or topographic changes that were not present on earlier
photographs. Sites that are obviously clean fill for construction purposes are not included in this category nor are they
annotated. (Such sites may be identified by the relatively
rapid completion of activity followed by the appearance of a
highway, new building or structure on more recent photographs.)

L-Lagoons:
    Potential liquid waste disposal areas that are either
suggested by associated activity on the photographs or are known
to have existed. Standing water in borrow pits or quarries is
not generally placed in this category unless associated with
dumping.

J-Auto Junkyards and Salvage Areas:
    Such sites may contain significant surface disposal or
spills or oil, transmission and hydraulic fluids, or solvents.

S-Suspicious:
    Areas where unusual or unidentifiable activity has occurred
that is not readily recognizable. Sites are placed in this
category pending more complete analysis that will result in one
of the above designations or elimination.

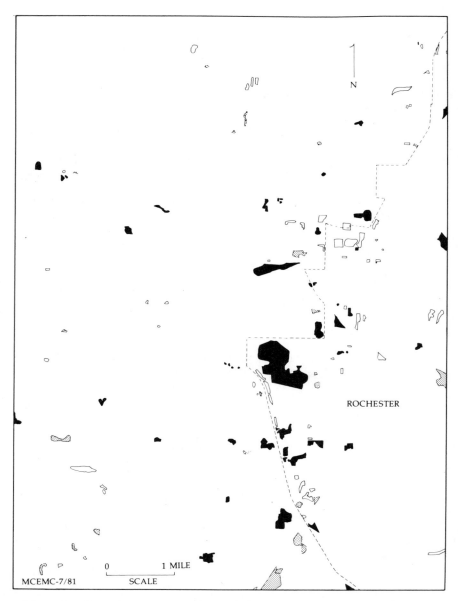

Figure 1.   General survey map of sites in the Rochester West quadrangle.   Key:
■, *dump;* ▨, *possible dump;* □, *unspecified fill.   (Reproduced, with permission,*
from Ref. 1.)

those factors which directly influence the production, containment, attenuation or migration of leachate. These generally
involve the groundwater system, the soil or rock permeability,
and the structures within the overburden or rock that control
either the direction of movement, rate of movement, or local
concentration of fluids. In most cases, landfills or old dumps
are located in unconsolidated soils or overburden but occasionally the character of the local bedrock is also significant.
Any geologist working on a project such as this needs to evaluate
the critical factors within the particular region under study.

   When dealing with abandoned landfills that were started or
completed prior to the late 1970's, it should be assumed that
wastes were generally placed in natural or man-made depressions
without utilization of sophisticated preparation or containment
techniques. Thus, the properties and structures of the natural
overburden and rock formations should be assumed to control the
natural migration of leachate and groundwater adjacent to the
site. While the Monroe County study emphasizes the problems
encountered in the glacial overburden deposits bordering Lake
Ontario, the basic approach should be applicable, with minor
modifications, to any regional analysis of hydrogeologic conditions.

   The first phase of the general geologic analysis consists
of the collection of all readily available published or unpublished data for the construction of regional geologic maps such
as those shown in Figures 2, 3, and 4. The collection of additional unpublished data should continue throughout the period of
study because useful or unique information can continue to be
located. The types of sources commonly used in the development
of a geologic data base for both general or site-specific geologic models are:

. Geologic maps and reports: published literature and student
  theses/dissertations; open-file reports of agencies such as
  United States Geological Survey; government documents.
. County soils maps: old and new versions of United States
  Department of Agriculture maps describing soil types and loca-
  tions of dumps or "made land" (good for shallow depth of
  overburden only).
. Aerial photographs and topographic maps: used for geomorpho-
  logical analysis of landforms if need to supplement or confirm
  published data. Compare old and recent prints or maps, if
  available.
. Engineering data from public or private agencies or firms
  (especially exploratory boring logs) -
  Highway/bridge plans or surveys.
  Town, village or county construction/maintenance projects.
  Utilities, railroads, pipeline companies, canals.
  Drilling and soil testing from engineering firms.
  Architectural firms.
  Oil, gas and water well drillers.

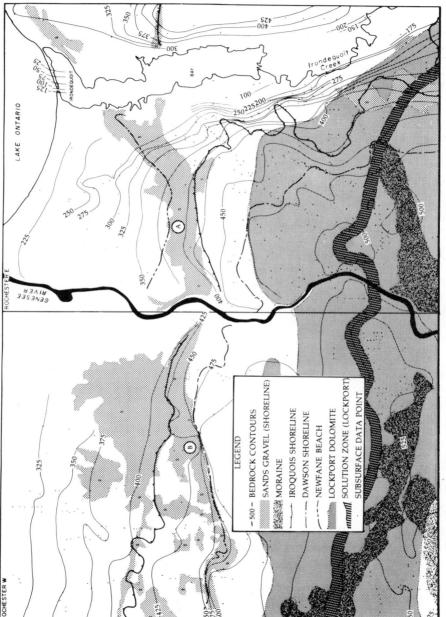

Figure 2. Subsurface bedrock contour map with selected surface geology for Rochester East and Rochester West 7.5-min topographic quadrangles. Contour interval is 25 ft. Map width is 12.6 mi.

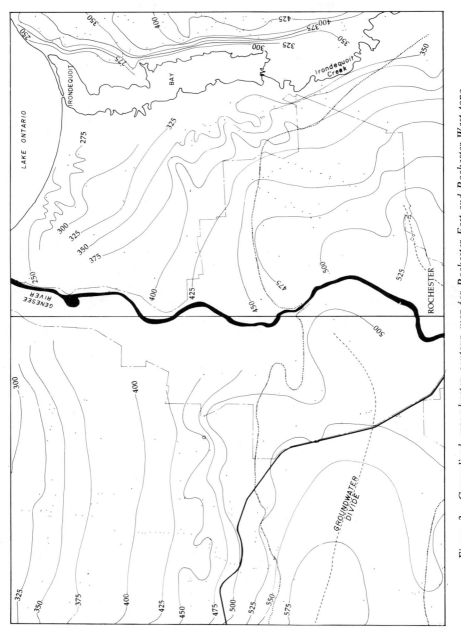

Figure 3. Generalized groundwater contour map for Rochester East and Rochester West topo-graphic quadrangles. The 25-ft contour intervals show water elevations in overburden and may reflect some seasonal variations. Map width is 12.6 mi.

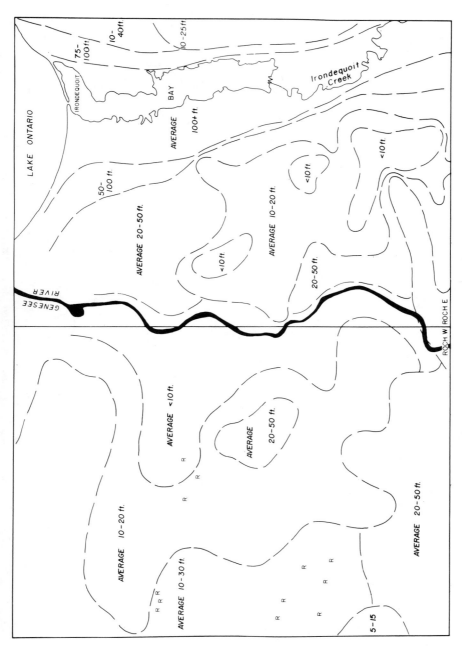

Figure 4. Overburden thickness map for Rochester East and Rochester West topographic quadrangles from same data points as Figure 2. R designates rock outcrops. Map width is 12.6 mi.

Sewer and water district agencies.
Mining, quarrying and tunneling ventures.
Public buildings, airfields and large industrial firms.

Geologic Overlay Maps. Using all available sources, maps such as those in Figures 2, 3, and 4 are constructed. It is recommended that a common scale be adopted for all such general survey map products and that they be compiled on transparent overlays so that all types of information can be easily transposed onto or compared with the general survey site maps. Topographic maps with a scale of 1:24,000 are very convenient for this purpose as much geologic data are routinely published on this scale by state and federal agencies.

These maps and the data from which they are generated have the following uses:
. Development and improved understanding of a regional groundwater model that includes depths to the water table and flow directions.
. Development of a regional perspective concerning areas where landfill or leachate problems would either be reduced or aggravated by local geologic conditions.
. Development of a basis upon which to prioritize or organize the search for sites with potential for significant impact.
. Allow an agency to focus its limited resources in the most critical areas.
. Extrapolation of subsurface conditions to sites where no data can be located from similar ones where detailed information exists.

Geologic Ranking Process. Once all the general maps have been prepared and subsurface engineering information compiled, the geologic ranking process can be applied using the Geologic Ranking Sheet (Figure 5). Table II explains the categories on the Geologic Ranking Sheet.

This ranking sheet has been devised for the following reasons:
. To provide a means of organizing the process of site comparison.
. To minimize inconsistencies and oversights that could occur when dealing with large amounts of data.
. To provide a record for other project personnel that can be updated and will allow discussion or review of site characteristics.

A ranking scheme such as this is basically an information gathering and documentation device that would be especially useful if project personnel changed.

Strict numerical site rankings should be considered only as approximations of site characteristics and geologic expertise should be substituted where appropriate. The system described here requires knowledgeable decisions to be made for each of the "presumed effects" and the ranking process should be done by an individual with hydrogeologic background.

GEOLOGIC RANKING SHEET
FOR GENERAL COMPARISON OF ABANDONED LANDFILL/DUMP SITES

SYMBOLS USED IN COLUMNS
X PROBABLE EFFECT
U UNCERTAIN: LIKELY EFFECT
⊗ EFFECT OF OVERRIDING SIGNIFICANCE
   Superscripts refer to footnotes.

SITE NAME/NO._____

SITE RANK
(CHECK ONE)

	PRESUMED EFFECT		
	A	B	C
FACTORS TO BE EVALUATED	HIGHER HAZARD	INTERMEDIATE (UNCERTAIN)	LOWER HAZARD
OVERBURDEN GEOLOGY [2]			
ESTIMATED PERMEABILITY [3]			
RELIEF, GEOMORPHOLOGY [4]			
SEPARATION OF WASTE FROM GROUNDWATER [5]			
GROUNDWATER GRADIENT [6]			
BEDROCK CHARACTER [7]			
SOIL MINERALOGY; TEXTURES [8]			
NUMBER OF ENTRIES			
MULTIPLIER	3	2	1
ENTRIES X MULITPLIER			

SUBTOTAL _____
ADDITIONAL FACTORS _____
TOTAL POINTS: SITE RANK _____

SITE RANK (CHECK ONE)	
.01 HIGHEST PRIORITY (17-21 PTS)	
.02 INTERMEDIATE PRIORITY (12-16 PTS)	
.03 LOWEST PRIORITY (7-11 PTS)	

_____ TENTATIVE   _____ FINAL

NOTE: IN CASES WHERE MORE THAN HALF THE CRITICAL FACTORS MUST BE RATED AS UNCERTAIN (U), THE RANK SHOULD BE TENTATIVE.

ADDITIONAL FACTORS
(CIRCLE AND ADD TO CHART)

THESE POINTS MAY INCREASE (+1),
DECREASE (−1),
OR NOT AFFECT (0) SCORE

VERY LARGE SITES (20+ ACRES)	+1
ENGINEERING/GEOLOGIC DATA ON OR NEAR SITE	0, −1, +1
GEOLOGY EXTRAPOLATED CONFIDENTLY FROM NEARBY	0, −1, +1

DESCRIBE IMPORTANT OR OVERRIDING FACTORS BELOW IF APPROPRIATE (DESCRIBE SPECIAL CONDITIONS): _____

_____
_____
_____
_____
_____
_____

Figure 5. Geologic ranking sheet. This form serves as a general guide for evaluating and recording geologic information on specific sites. See Table II for explanation.

Table II.

Description of Factors on Geologic Ranking Sheet

1. <u>PRESUMED EFFECT</u>: A decision is required as to whether each
    inferred or documented FACTOR would increase or decrease the
    hazard relative to leachate production, migration, or atten-
    uation. No simple, uniform guidelines can be set forth that
    cover all situations or geohydrologic complexities.

2. <u>OVERBURDEN GEOLOGY</u>: From inferred nature of unconsolidated
    sediments would leachate occurrence be likely to increase or
    decrease human exposure to pollutants?

3. <u>ESTIMATED PERMEABILITY</u>: Is estimated permeability of uncon-
    solidated materials likely to increase or decrease exposure
    risks? Include estimated effect of either aquifers or aqui-
    cludes or inferred combinations.

4. <u>RELIEF, GEOMORPHOLOGY</u>: Does relief on or adjacent to the
    site influence the occurrence or migration of leachate so as
    to increase or decrease the exposure hazard?

5. <u>SEPARATION OF WASTE FROM GROUNDWATER</u>: Does the estimated
    depth to the water table imply a high or low risk for con-
    tamination or leachate production. Relate to permeability
    and gradient factors.

6. <u>GROUNDWATER GRADIENT</u>: Gradient is dependent on local relief,
    estimated permeability, aquifer characteristics and rainfall
    patterns. Steep or flat gradients by themselves cannot be
    presumed to have similar effects in each case. Judgment is
    required on local conditions.

7. <u>BEDROCK CHARACTER</u>: Is local bedrock an important factor in
    local hydrologic system? If so, do textures or structures
    in bedrock produce asymmetry or enhanced flow of potential
    leachate plume (flow along bedding, joints, faults, or
    solution channels).

8. <u>SOIL PROPERTIES, TEXTURE AND BEHAVIOR</u>: Are there known tex-
    tural or mineralogical factors that could enhance or diminish
    leachate migration, such as strong cation exchange or swell-
    ing/shrinking clays (cracking)? Are seasonal effects such as
    rainfall duration, infiltration capacity, freeze-thaw condi-
    tions, vegetation cover, etc. of significance?

The ranking sheet was designed to divide potential sites into high, intermediate or low priorities based on anticipated geologic conditions governing leachate (groundwater) occurrence, production, migration or accumulation, but it is only one component of an integrated assessment system. Its prime consideration is the potential or inferred effects that leachate contamination would have on people near the sites.

High priority sites are those where hazardous materials might accumulate, move readily at shallow depths, or reappear at the surface in concentrated amounts. Low priority sites are those adjacent to lakes, ravines, or on steep slopes and presumed to have been well drained for an extended period. Any large site with known hazardous contents should be evaluated individually and referred to the appropriate agency regardless of its position in the geologic ranking system.

Intermediate sites are those falling between the extremes described above. They would generally be sites where geologic conditions contain and isolate any leachate or partially attenuate its effects.

Potential water well contamination involves geologic considerations but sites near public or private drinking water supplies are treated separately due to the potential direct risks involved.

Use of the Ranking Sheet. Each factor on the ranking sheet is analyzed as to whether it increases or decreases the potential risk for a site. Available engineering information and potential interrelations among factors need to be considered. When the sheet subtotal is computed, the "additional factors" are evaluated and a total score computed. It is important that borderline cases be carefully evaluated and a tentative rank assigned if necessary, pending further data collection. Special problems should be noted in the space provided at the bottom of the sheet.

This system is designed to deal with a large number of abandoned sites where information is scarce or difficult to locate. It is not meant to replace the elaborate systems for ranking known hazardous waste sites such as those by LeGrand[2] and Kufs et al.[3] The Monroe County study is intended to broadly prioritize sites, rapidly identify potentially high risk sites and allow the most efficient use of funds available for later testing or subsurface investigations.

Refinement of Site Information

Upon completion of the site identification phase, there will be sites categorized as "Possible Dumps" or "Unspecified Fills." Additional information must be obtained in order to clarify which of these sites were actually used as dumps. Information can be obtained from existing agency records, interviews with

local officials, waste haulers, industrial employees, historic
documents and street and business directories.

   Table III indicates the number of dumps identified for one
town in Monroe County before and after contact with local officials and field inspections. The findings illustrate that the
methodology does not overestimate the number of dumps and possible dumps identified through photo interpretation. However,
the process was able to identify approximately twice the number
of dumps initially identified through a call-in campaign and
preliminary interviews with the same local officials (10).

Table III.   Categorization of Sites, Greece, New York

	Dumps	Possible Dumps	Unspecified Fills	Total
After Photo Interpretation	11	19	11	41
After Final Interviews with Local Officials	19	13	6	38*
After Field Inspection	21	11	6	38*

*Three sites were eliminated because they were found to be clean
 fill for construction or landscaping.

   Once the site categories are refined, research is conducted
to determine the location of public and private drinking water
supplies in proximity to known dump sites. Where private wells
are found within 1,000 feet of a known dump, or public water
wells within one-half mile, geologic conditions are carefully
evaluated to determine if contamination could occur. When this
determination is positive, the sites are referred to the appropriate local or state agency for water well testing.

Site Prioritization

   All dump sites are now prioritized in a matrix using
geology and land use on the two axes. Table IV shows the final
ranking of dump sites for the town of Greece.

Table IV. Prioritization Matrix for Waste
Disposal Sites, Town of Greece, New York

Land Use Categories

		.01	.02	.03
	.01	4	0	0
Geologic				
Categoies	.02	3	5	0
	.03	3	5	1

Note: The sites classified as .01/.01 represent the highest
priority in terms of potential impact.

Summary and Conclusions

In the final analysis, no general program for identifying
abandoned dump sites can ever completely determine the location
and contents of all hazardous waste disposal areas. The Monroe
County approach is designed to provide general information on a
large number of previously undocumented sites, as well as a
method for local, state and federal agencies to prioritize sites
for more detailed site investigation and testing. Application
of the procedures will reduce the expense of costly drilling and
testing programs by focusing resources on the most critical
sites.

Acknowledgments

Major portions of this article were printed in the
Proceedings of the National Conference on Management of Uncontrolled Hazardous Waste Sites, October 1981(1) and are reprinted
with permission of the Hazardous Materials Control Research
Institute. This text, however, has been revised to reflect
further refinements in the methodology that have occurred since
the original publication.

This study has been conducted under the direction of the
Monroe County Landfill Review Committee, comprised of representatives of the Monroe County Environmental Management Council,
the Monroe County Departments of Health and Planning, the New
York State Departments of Health and Environmental Conservation,
the City of Rochester Department of Community Development and
the Industrial Management Council.

The development of the methodology has been made possible
through the financial support of the New York State Department
of Environmental Conservation in 1979 and 1980, and currently
through the United States Environmental Protection Agency's

Environmental Monitoring Systems Lab in Las Vegas, Nevada (Contract No. 14043 administered by Lockheed Engineering and Management Services Company, Inc.).

Literature Cited

1. Nelson, A. B., Young, R. A. *Proc. Nat'l. Conf. on Management of Uncontrolled Hazardous Waste Sites* 1981, p. 52.
2. Legrand, H. E. *A Standardized System for Evaluating Waste Disposal Sites:* A manual to accompany description and rating charts; National Water Well Association 1980.
3. Kufs, C., Twedell, D., Paige, S., Wetzel, R., Spooner, P., and Colonna, R. *Proc. U. S. EPA National Conf. on Management of Uncontrolled Hazardous Waste Sites* 1980, p. 31.

RECEIVED June 16, 1982.

Monitoring to Support Risk Assessments at Hazardous Waste Sites

GLENN E. SCHWEITZER

U.S. Environmental Protection Agency, Environmental Monitoring Systems Laboratory, Las Vegas, NV 89114

The classical approaches to multimedia monitoring at hazardous waste sites are briefly reviewed, with examples from recent site investigations. The types of monitoring data that are most useful in risk assessments are discussed in terms of feasibility, costs, and time for acquisition. Required emphases for assessing environmental hot spots and for determining long-term habitability of a larger area are described. Special attention is directed to geophysical techniques, biological monitoring, and the triple stage quadrupole mass spectrometer.

In recent years environmental monitoring activities have been undertaken at hundreds of hazardous waste sites throughout the country. While some of these efforts date back a decade or more, most of the monitoring has been initiated during the past several years in anticipation of or in response to Governmental regulatory actions at the national, state, and local levels.

Often the monitoring has been targeted on one, two, or a small number of chemicals. In other cases the 129 priority pollutants have been the subject of investigations. In a few cases a variety of analytical methods have been used in efforts to identify and measure still larger numbers of possible environmental containments.

As indicated in Figure 1, all environmental media where pollutants could lodge have been of interest, including air, surface water and sediment, soil, and groundwater. The environmental pathways of greatest concern obviously vary from site to site. Figure 2 summarizes the media of principal interest at nine "typical" sites selected from an EPA survey of 160 sites. In addition, man-made pathways for pollutants have taken on new significance -- sewage and drinking water systems, irrigation and drainage systems, and cut and fill activities. Even measurements

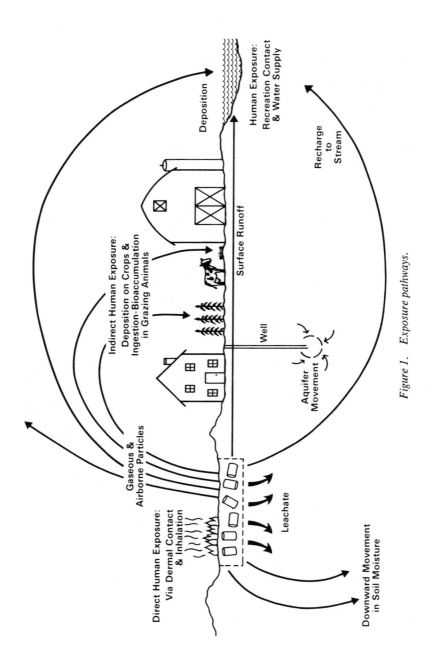

Figure 1. Exposure pathways.

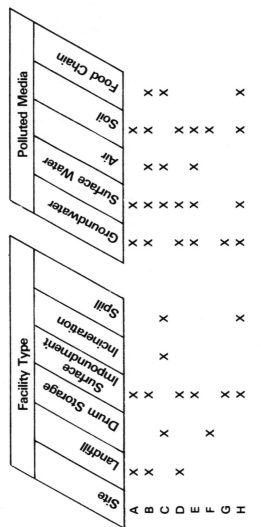

Figure 2.   Case studies of typical inactive sites.   (Reproduced from Ref. 1.)

of indoor air quality and basement sump contamination have been
linked to waste site problems.

Remote monitoring techniques have been used to complement
conventional environmental monitoring activities. The use of
current and archival aerial photography and multispectral over-
head imagery to uncover and delineate waste site problems has
expanded considerably. Also, ground penetrating remote sensing
technologies are being used to delineate the subsurface distribution of waste materials and to help target monitoring activities. In addition, these geophysical technologies have been
shown to offer promise for detecting underground movement of
pollution plumes.

Biological monitoring has occasionally been conducted at
hazardous waste sites to clarify possible food chain problems.
Food crops, livestock, and local fish in particular have been of
concern. In addition, wild animals and indigenous vegetation
have been sampled to find indications of local contamination
problems.

In short, the large number of multi-chemical, multi-media
environmental measurements associated with hazardous waste
investigations during the past few years are unprecedented in
scope and complexity. The small community of specialists in the
measurements field is being pushed to the limit to reduce the
time and cost involved in such measurements and to improve the
representativeness of the data that are collected and the
accuracy of the data that are reported.

Relating Monitoring Data to Risks

Monitoring activities have usually been carried out to help
clarify the risks, if any, associated with specific sites. While
every data point that could be obtained may be of some interest,
the challenge is to maximize the usefulness for risk assessment
of monitoring data that are collected and analyzed within cost
and time constraints. Unfortunately, health assessment specialists have provided little guidance to date as to the type of
monitoring information that would be most useful to them in
determining risks to individuals or to larger populations.

One approach to relating monitoring at a hazardous waste
site to risk assessment is simply to determine whether existing
air or water quality standards are being exceeded in the vicinity
of the site and whether such excesses are due to the presence of
the site. With regard to groundwater, Maximum Contaminant Levels
(MCLs) established for drinking water have been suggested as
surrogate standards. A principal drawback in this approach is
that standards and MCLs exist for only a few chemicals. Another
obvious deficiency is that when a standard has been exceeded, it
may be too late for effective remedial action.

A second type of approach is to compare contamination levels
in the media of interest near the site with pollution levels in

the same media in control areas. Of special interest would be a
control area near the waste site--but insulated by geographic
features from the direct influence of the site. This area would
be impacted by all of the industrial emissions and effluents that
permeate the region as well as other common sources of contamination such as agricultural chemicals. Thus, in comparing the
contamination near the site with the contamination in the control
area, it should be possible to attribute any higher levels of
contaminants found near the site to the influence of the site
itself and not to the background characteristics of the region.
A second type of control could be provided by national baseline
data. Such data indicating the levels of ambient or background
contamination usually encountered in different types of demo-
graphic settings could help clarify the significance of environmental measurements near a hazardous waste site.

The classical approach of estimating multimedia exposures to
individual chemicals and then estimating the related risks to
sensitive populations through a combination of modeling, monitoring, and toxicological evaluation does not appear very promising.
Usually there are too many chemicals and chemical mixtures
involved, too many uncertainties in estimating chemical fate and
transport and hence the risks to future generations, and too many
unknown concerning future site contaminant failures to give much
credence to such approaches. Still, in cases involving highly
toxic chemicals such as dioxin contamination, detailed analysis
of the presence and likely environmental behavior of such individual chemicals may be important, and monitoring capabilities to
detect and measure the presence of such chemicals at low concentrations will be critical.

Finally, several ranking systems have been developed for
prioritizing the relative hazards at a large number of sites.
This approach usually involves a weighting of the amount and
composition of hazardous material at each site, the contaminant
characteristics of the site, the potential migration pathways,
and the proximity of environmental or human populations. Variations of this approach have already been applied many times,
either in a quasi-rigorous fashion or in a more general manner,
as state and local authorities are faced with deciding where to
concentrate limited resources.

The General Approach to Sampling and Analysis

Monitoring activities involve several interrelated steps:
-- Deciding where and how to take representative samples of
the media of interest and how to handle the samples en route to
the analytical laboratories.
-- Selecting analytical methods to investigate the presence
or absence of either a finite or open ended list of chemicals.

-- Choosing a procedure and format for aggregating and presenting the results of the analysis and for setting forth the degree of confidence in the data.

-- Determining how to interpret the monitoring data as to presence, quantity, transformation, and migration of the chemicals of interest.

We are rapidly acquiring experience in all of these areas. Over the years many manuals have been prepared by a number of groups on sampling and analytical methods. Sampling design is usually site-specific, and sometimes chemical-specific, and case studies of different approaches that have been used in the past are particularly instructive in providing future guidance in this area. However, with regard to the presentation and interpretation of data, we have only begun to develop useful approaches for relating monitoring data to risks--the thrust of this symposium.

A few of the lessons we have learned to date in carrying out monitoring programs are as follows:

1. At the outset the objectives of the monitoring program should be clearly defined. Monitoring programs can be helpful both in identifying hot spots of near term concern and in assessing the longer term habitability of the area. Each of these objectives may require a different program orientation, however, as shown in Figure 3. Usually both objectives will be important in varying degrees, and the program emphasis should be adjusted accordingly. Prior to initiating the monitoring program, the users of the monitoring data should be satisfied that the data will be collected, analyzed, and presented in a manner which is responsive to their needs.

2. Several factors are important in selecting the individual chemicals of special interest. For example, certain chemicals may have been deposited at the site in very large quantities, and therefore they are of principal concern simply because of their volume. Secondly, one or more of the chemicals known to be present at the site may be so toxic that the potential threat even at low volumes is obviously a major concern (e.g., dioxin). Certain "indicator" chemicals commonly found at waste sites may behave in the environment in a manner characteristic of many other chemicals as well (e.g. groups of pesticides), and determining the environmental distribution of such "indicator" chemicals could be indicative of broader contamination problems. Analysis of leachate from the edge of the site or preliminary monitoring close to the site might identify chemicals that are escaping from the site. Finally, sampling for a group of chemicals which are usually analyzed as a package such as the 129 priority pollutants might result in considerable savings per chemical in analytical costs. All of these factors should have a bearing on the selection of the chemicals for intensive investigation.

3. Statistical aspects are an important factor in the design of a monitoring program that is to provide representative

	Assessing Hot Spots	Assessing Chemical Saturation
Objective	Identify near term exposure problems	Determine long term habitability of area
Chemicals of concern	High toxicity chemicals suspected to be in area	Many chemicals including degradation products
Levels of concern	"High" levels	All levels including trace levels
Likely sources	Past and current leakages at site	Past, current, and future leakages
Pathways of primary concern	Air, drinking water, food	Also, soil, sediment, biota, surface water, sewers, groundwater
Populations of concern	Susceptible individuals near hot spots	Entire population
Assessment emphasis	Determine highest intergrated exposure levels for specific groups of individuals	Document highest and median concentration levels in individual media and compare such levels with control areas
Monitoring approach	Limited multimedia sampling of broad areas and intensive single media monitoring of suspected hot spots near populations. Emphasize monitoring at interface of receptor.	General multimedia sampling of broad areas with bias toward natural and man-made pathways from the site. Concurrent monitoring in nearby control areas. Monitor along entire environmental pathway.

Figure 3.   Assessing exposures near uncontrolled sites.

data. A statistician on the planning team can help insure that
adequate consideration is given to these aspects both in designing the program and in formatting and interpreting data. A photo
interpreter can also be an important member of the planning team,
both in selecting monitoring sites and in designing the approach
for relating concentration levels to population activities.

4. Before any samples are taken, a complete sampling plan
should be developed with built-in check points for adjusting the
plan as sampling results are obtained. Deviations from the plan
should be resisted other than to adjust the plan in response to
unexpected data results. As a rule of thumb, on the order of
fifteen percent of the sampling might be for screening to help
clarify hypotheses as to possible gradients and hot spots around
the sites, seventy percent directed to investigating the
hypotheses, and fifteen percent reserved for supplementary
sampling of neglected areas that come to light late in the
program. The sampling scheme should include provisions to
confirm or reject previously reported findings of a controversial nature. Access to preferred sampling sites is not always
possible. The sampling plan should be sufficiently flexible to
compensate for such problems.

5. A quality assurance program involving surrogate recov-
eries, inter and intra laboratory duplicates, and field and
laboratory blanks is essential. Each data point should be
individually validated as acceptable data, and precision and
accuracy data should be developed for each data set. The quality
assurance program may account for 10 to 20 percent of the monitoring costs. Special efforts are needed to minimize holding
times between sampling and analysis. However, extended holding
times beyond the usual target of two weeks may be unavoidable.
In that event appropriate storage procedures are particularly
important to prevent excessive decay of the samples. Also,
contaminants associated with the sampling and analytical
techniques are often difficult to avoid, and data suspected
of such contamination should be considerd for discarding. Of
particular concern, for example, are benzene and toluene
contamination when using Tenax traps, methylene chloride and
phthalates that are present in laboratories, isophorone which can
be a derivative of the laboratory solvent acetone, and the high
pH in groundwater associated with grouting of sampling wells
which may result in artifacts being observed.

6. Data formatting and presentation can have a significant
impact on the interpretation of the data. Plotting each data
point on a map is probably the safest way to insure a totally
objective presentation of findings. Monitoring data may not
provide a definite portrayal of pollutant gradients or pollution
patterns but may only be suggestive of general pollutant
distribution. Interpretations of the data may be controversial,
and efforts should be made to isolate criticisms of the interpretations.

These and other lessons learned set an important framework
for our research efforts to improve monitoring capabilities.
Discussed below are three areas of current research interest.

Geophysical Techniques

The increasing importance of geophysical investigations has
been repeatedly documented in recent months. The most commonly
used techniques have been seismic refraction, ground penetrating
radar, electrical conductivity, and magnetometer surveys. Figure
4 presents a comparison of an analysis of data from 16 groundwater sampling wells with an analysis of electromagnetic data in
determining plume flow. Figure 5 underscores how conductivity
data might have been used to improve the location of sampling
wells. Finally, Figure 6 reflects the potential contributions of
geophysical investigations to help plan remedial actions.
Efforts have also been initiated to instrument new sites
with electrode systems for detecting leachate leakage through the
liners of the sites. One approach is to surround the site with
resistivity sounding stations using the earth as the conducting
medium. Another approach has been to embed a wire grid just
below the site. In either case, changes in resistivity measurements would signal a possible leachate plume migrating downward
from the site. While this technique seems very appealing, considerable proof testing is in order given the necessity for systems that will minimize false alarms and will operate reliably
for many decades.
Another promising approach calls for combining laser-induced
fluorescence to indicate pollutant contamination of an aqueous
body, and in this case groundwater, with fiber optic techniques
for entering the earth through very small diameter wells (less
than one inch). This approach is particularly attractive in
attempting to measure Total Organic Carbon (TOC) as a surrogate
for the pollution plume. Indeed, a principal research direction
is demonstrating the feasibility of TOC measurements while
searching for additional pollution signatures that could be
detected by laser fluorescence. If the method proves opera-
tionally feasible, reductions in well diameters should result in
considerable cost savings.

Improved Analytical Methods

As more potentially harmful chemicals are discovered,
concern for the long-term impact of low concentrations of large
numbers of hazardous elements and compounds increases. The need
for improved methods for analyzing environmental samples to
identify and measure very small amounts of a wide variety of
chemicals becomes critical.
A number of advanced methods for analyzing environmental samples are under investigation by researchers in many

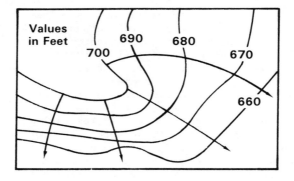

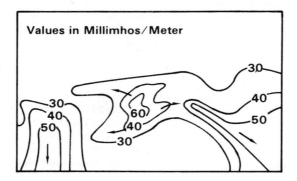

Figure 4. Monitor well measurements and electromagnetic measurements at same site. Top: potentiometric contours showing plume flow. Bottom: conductivity contours showing plume flow. (Reproduced from Ref. 2.)

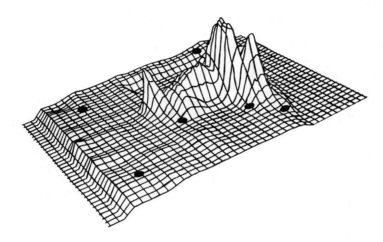

Figure 5.   Conductivity data showing buried hazardous materials. Key: ●, wells drilled before conductivity data were available. (Reproduced from Ref. 2.)

Alternative	Informational Needs
Removal of Buried Drums and Chemicals	**Drum Condition, Number, and Contents** **Trench Location and Geometry**
Encapsulation of Source	**Drum Contents** **Imperviousness of Underlying Strata** **Level of Subsurface Contamination** **Trench Location and Geometry**
Collection and Treatment of Leachate	**Areal Extent of Contamination** **Type of Contamination** **Concentration of Contaminants** **Imperviousness of Underlying Strata** **Aquifer Characteristics**
No Action	**Drum Contents and Condition** **Level of Subsurface Contamination** **Type of Contamination**

Figure 6.   Informational needs for implementation of abatement activities. (Reproduced from Ref. 1.)

laboratories. Of particular relevance to the problems of
hazardous wastes are the application of fused silica capillary
columns in GC/MS systems, Fourier transform infrared spectros-
copy, and non-extractive multi-elemental analysis.

Present MS techniques for analysis of organic compounds
in complex samples require separation of the sample components
prior to MS analysis. The separation is usually accomplished by
extracting the samples, separating the extracts into several
fractions (cleanup), and analyzing the fractions by GC/MS. These
steps, especially the sample preparation and cleanup, are time-
consuming and expensive.

The use of tandem mass spectrometers can eliminate the
sample preparation steps and provide improved capabilities for MS
analysis. One system, the triple quadrupole mass spectrometer,
uses a combination of three quadrupoles, or mass analyzers, to
ionize, separate, and analyze sample components with minimum
sample preparation as shown in Figures 7 and 8. The sample
components are ionized and separated according to their mass-
to-charge ratio in the first quadrupole. This step corresponds
to the GC step in GC/MS. In the second quadrupole these ions
collide with an inert gas and fragment (chemical ionization).
In the third quadrupole the fragments are identified (mass
analyzer).

Triple quadrupole mass spectrometry can provide rapid
screening of complex mixtures for specific compounds and can be
used to analyze for compounds that cannot routinely be analyzed
by GC/MS. In addition, structural information can be obtained
for certain types of compounds since in collision-induced
dissociations the fragments are likely to show the structual
differences of the parent compounds. Complex mixtures have
been analyzed by this technique by introducing the sample
directly into the heated sample port of the instrument with
little or no sample pretreatment. Triple quadrupole mass
spectrometry promises to be a useful, cost-effective, and
practical advanced technique for environmental analysis,
particularly when applied to hazardous waste problems.

Monitoring in Animals and Plants

As indicated in Figure 9, many approaches to monitoring
animal species at hazardous waste sites have been suggested.
Also, vegetation monitoring is commonly proposed. Very few of
the suggested approaches, however, are beyond the early research
stage to the point where they could be deployed to provide
reliable data concerning field contamination.

The types of biological methods that need to be carefully
evaluated and, if appropriate, standardized for field use include
(1) laboratory screening tests to assess the relative hazard of
chemical mixtures, (2) field survey methods to detect changes in
biological populations, and (3) monitoring techniques to detect

Figure 7.   Triple-stage quadrupole mass spectrometer in operation.

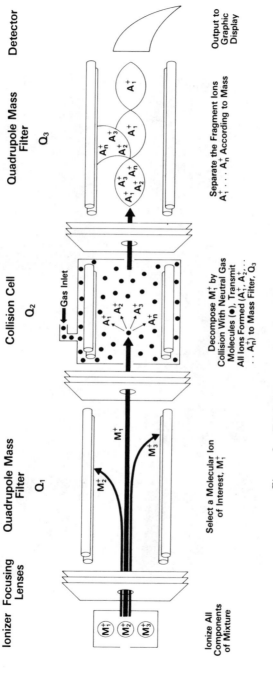

Figure 8.   Triple-stage quadrupole mass spectrometer schematic.

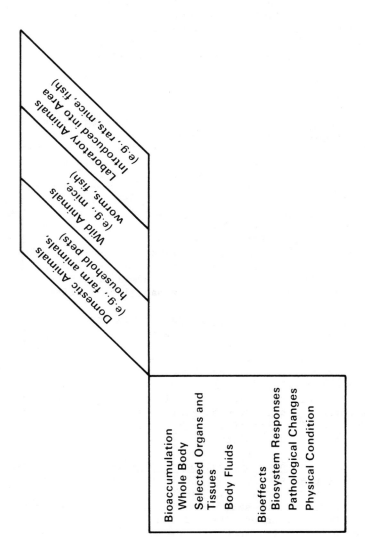

Figure 9.   Biomonitoring using selected animal species.

the presence of specific compounds in biological materials. The
three approaches have several features in common. These methods,
when applied within appropriate experimental designs, may indicate the source of pollutants and potentially can indicate
pollutant distribution in the air, water, soil, or sediment at
study locations. Equally important, they can identify the
extent of problems, they can provide information useful for
hazard evaluations, and they can frequently help identify
specific pollutants, or combinations of pollutants, that are
causing problems.

Screening Tests. Screening tests can be used during initial
investigations at waste sites to provide a rapid assessment of
potential hazards to (1) field and laboratory personnel investigating the site, (2) persons in the immediate vicinity of the
site, and (3) the nearby environment. Exposures of the screening
systems are conducted under controlled conditions, usually by
bringing samples into the laboratory or by placing test organisms
at the site. Examples of screening tests to clarify relative
personnel hazards might include the detection of mutagens in body
fluids of laboratory animals exposed to waste site material and
various immunotoxicity responses in laboratory animals. The
development of biological screening tests has received much
attention over the past few years, and several tests (e.g., the
Ames test) have become widely used by many laboratories.

Surveys of Environmental Impacts. Field survey methods
focus on assessing changes in the condition of entire populations
and ecological community functions (e.g., pest species prolif-
eration, species diversity, litter decomposition rate, changes in
the rate of primary production). These methods under certain
circumstances may relate effects to specific pollutant sources,
but most frequently they are used to indicate that a problem does
exist. A current application of survey methods being developed
emphasizes identification of indigenous species to serve as
surrogates for a portion of the ecosystem.

Monitoring for Bioaccumulation or Bioeffects. Biological
methods can be used simply to indicate the presence of a
pollutant at a study location. Chemical analysis can identify
the type of pollutant present in biological species as shown in
Figure 10. However, unless there is an extensive data base
concerning uptake rates, such an approach will not reveal
pollutant concentrations in the area. Biological effects such
as enzyme inhibition can also be used in some cases to indicate
the types of chemicals in the area (e.g., inhibition of red blood
cell acetylcholinesterase following an animal's contact with
organophosphorus chemicals). The detection of metabolic products
in body fluids might also be used to investigate the biological

	Mammals, birds, and fish	Mollusks, crustacea, and lower animals	Higher plants	Mosses, lichens, and algae
Antimony	x	x	x	x
Arsenic	xx	xxx	xxx	xx
Beryllium	x	x	x	x
Boron	x	xx	xxx	x
Cadmium	xxx	xxx	xxx	xxx
Chromium	xx	xxx	xxx	xxx
Cobalt	x	x	xxx	x
Copper	xxx	xxx	xxx	xxx
Lead	xxx	xxx	xxx	xxx
Mercury	xxx	xxx	x	x
Nickel	xx	xxx	xxx	xxx
Selenium	xx	x	xxx	x
Tin	xx	x	xxx	x
Vanadium	x	xxx	xx	xx

x—Low or limited.   xx—Moderate.   xxx—High to very high.

Figure 10. Bioaccumulation and bioconcentration of trace elements.

uptake of some compounds (e.g., the presence of chlorinated phenols in urine following uptake of chlorinated hydrocarbons).

Providing More Meaningful Monitoring Data for Health Assessments

Even with rapid advances in the state-of-the-art of monitoring technologies many of the problems of effectively meshing monitoring data with data on health effects will continue to persist. Undoubtedly, the past approaches of colocating monitoring and health studies will continue. These approaches rely on monitoring activities to clarify exposures to specific chemicals and on health studies either to assess the accumulation of these chemicals in human populations or to clarify the health effects which could be reasonably attributed to chemical exposures. Good examples of these two approaches are the efforts in the United States and abroad to (a) relate blood levels of cadmium and lead both to exposures and to effects and (b) relate certain types of exposures to organic chemicals to increased incidence of cancer. Obviously, uncertainties abound in such efforts particularly with regard to characterizing population activity patterns over an extended period of time as the basis for exposure estimates.

Turning more directly to the issues surrounding hazardous waste sites, the scientific community has an unusual opportunity to develop and apply improved approaches for coupling environmental chemistry and the health sciences. While answers may be needed today, the same types of answers will be needed for many decades. Thus, we should think in terms of long-range research programs that will provide a variety of interim answers--programs that can relate measurements in soil, water, air, and the food chain in meaningful ways to near-term and long-term effects in people. For example, how can monitoring help us predict with some confidence the rates at which chemicals under a variety of conditions will migrate toward people? Can aggregations of data on classes of chemicals detected at trace levels provide a meaningful basis for investigating possible effects? Can practical monitoring systems be developed which provide exposure data on individuals as is often done using film badges and thermoluminescent dosimeters in the radiation field? Finally, can measurements of very low levels of chemicals or of the early stages of biochemical reactions in people or animals serve as early warning indicators of adverse health effects that could result from further exposures?

Literature Cited

1. "Remedial Actions at Hazardous Waste Sites: Survey and Case Studies," United States Environmental Protection Agency 430/9-81-05.

2. "National Conference on Management of Uncontrolled Waste Sites," United States Environmental Protection Agency/ Hazardous Waste Material Control Research Institute, October 28, 1981, pp. 86, 87.

RECEIVED June 16, 1982.

Currently Available Geophysical Methods for Use in Hazardous Waste Site Investigations

ROY B. EVANS

U.S. Environmental Protection Agency, Environmental Monitoring
Systems Laboratory, Las Vegas, NV 89114

Monitoring to estimate exposure is essential to
risk assessment near sources of exposure. Cost-
effective hazardous waste site assessments include
three phases: (1) preliminary site assessment,
utilizing aerial photography and site inspections;
(2) geophysical surveys to pinpoint buried wastes
and to help define plumes of conductive contaminants in groundwater; and (3) confirmation of
groundwater contamination through monitoring well
networks designed on the basis of the geo physical
survey. Currently available geophysical methods
most applicable in hazardous waste site investigations include metal detectors and magnetometers
(useful in locating buried wastes); ground-
penetrating radar (useful in defining trench
boundaries); electromagnetic induction or EM
(useful in surveys of shallow plumes of conductive
groundwater contaminants); resistivity (useful in
surveys of site stratigraphy and deep groundwater
contaminant plumes); and seismic methods (most
useful in surveying geologic stratigraphy).

Monitoring is usually an essential part of risk assessment;
environmental samples are collected in various media near
sources of potential exposure, and these samples are analyzed
for toxic substances suspected of being present. In the past,
investigation of hazardous waste sites has commonly depended
upon drilling to obtain information on the geologic setting,
upon monitoring wells for samples of groundwater, and upon
laboratory analysis of soil and waste samples. During the past
decade, extensive development in remote sensing geophysical
equipment, field methods, analytical techniques, and associated
computer processing has greatly improved our ability to
characterize hazardous waste sites. Some geophysical methods
offer a direct means of detecting contaminant plumes, flow

direction, and buried drums. Some are applicable to
measurements of contaminants and direction of flow within the
vadose zone; others offer a way to obtain detailed information
about subsurface geology. The capability to characterize the
subsurface rapidly without disturbing the site offers benefits
in terms of less cost, less risk, and better understanding of
site conditions.

Cost-effective design of hazardous waste site groundwater
investigations involves an integrated, three-phased approach:
(1) preliminary site assessment, involving the use of aerial
photography, on-site inspections, and readily available
information to approximate site boundaries and locations of
waste concentrations, as well as probable site geology; (2)
geophysical surveys to pinpoint buried wastes and estimate
quantities, and to delineate plumes of conductive contaminants
in groundwater; and (3) confirmation of groundwater
contamination through monitoring well networks designed on the
basis of plumes and subsurface stratigraphy defined by the
geophysical surveys. The spatial characterization of the site
by geophysical means can make possible the efficient location of
monitoring wells and the reduction of risks involved in
exploratory drilling.

Field Problems

The three general objectives usually involved in subsurface
investigations are:

- Location of buried materials,
- Determination of the presence of plumes and the
  direction, rate of movement, and distribution of
  contaminants;
- Characterization of the geohydrologic conditions,
  natural and manmade.

Location of buried materials at a hazardous waste site is
usually for the purpose of remedial action; i.e., excavating
these materials and ultimately disposing of them. The key
unknowns are type (bulk-dumped or packaged in drums or other
containers), quantity (volume of waste; number of drums), and
location, particularly depth of burial. The concerns are for
safe excavation without puncturing containers or breaching any
existing trench liners and thus aggravating the cleanup
problems.

Determination of the presence of contaminant plumes and
their flow direction and movement rate is commonly required at
hazardous waste sites. The first determination is whether
leakage from the hazardous waste site is occurring. If the
existence of a plume is confirmed, its direction and extent
should be established and identified. A preliminary geophysical
survey can aid in better defining the contaminant plume, leading
to more effective monitoring with a smaller number of monitoring
wells.

   Characterization of the natural setting is usually a major
portion of the field investigation. At most sites, permeability
of the local soil and rock types, the depth of the water table,
and the direction of groundwater flow will strongly influence
movement of contaminants from the point of disposal. The
anomalies which occur naturally within the geohydrologic section
must be taken into consideration. Surface drainage, sewers, and
buried utilities can affect surface and groundwater flow around
a hazardous waste site.
   Many of these problems can often be avoided by the use of
an integrated approach combining contemporary geophysical
methods to support traditional drilling procedures.

Available Geophysical Methodology

   This paper discusses six currently available geophysical
sensing methods: ground penetrating radar (GPR),
electromagnetic induction (EM), resistivity, seismic refraction,
metal detection, and magnetometry. Other geophysical methods
are available and are proving to be effective; still others (for
example, complex resistivity) are emerging. However, this
discussion has been limited to the six methods involved in the
largest number of actual site investigations to date. Table I
shows the possible roles of each of these methods in hazardous
waste site assessments. These methods should be regarded as
complementing one another. No one method is used for all
applications, and some methods are useful in more than one
application. In general, metal detectors and magnetometers are
most useful in locating buried wastes; ground penetrating radar
is the technique of choice for defining the boundary of buried
trenches; electromagnetic induction (EM) and resistivity are the
most useful in defining plumes of conductive contaminants in
groundwater; and resistivity and seismic techniques are most
useful in determining geologic stratigraphy.

   Metal Detectors. Metal detectors respond to changes in
electrical conductivity caused by the presence of metallic
objects, both ferrous and non-ferrous. At the same time, metal
detectors are relatively insensitive to changes in soil moisture
or groundwater conductivity. The magnitude of response from a
metal detector is a function of several variables.
   1)   Target to sensor distance (response falls of as the
        sixth power of the distance)
   2)   Target size
   3)   Target orientation
   4)   Target geometry
   5)   Type of target metal
   6)   Mechanical/electrical integrity of the target
   7)   Search coil size
   The operation of simple metal detectors is diagrammed
schematically in Figure 1, which shows an induction balance

TABLE I.   POTENTIAL APPLICATIONS OF GEOPHYSICAL METHODS

	GROUND PENETRATING RADAR	ELECTROMAGNETIC INDUCTION	RESISTIVITY	SEISMIC	METAL DETECTOR	MAGNETOMETER
MAPPING OF CONDUCTIVE LEACHATES AND CONTAMINANT PLUMES (E.G., LANDFILLS, ACIDS, BASES)	(X)	X	X			
MAPPING OF GEOHYDROLOGIC FEATURES	X	X	X	X		
LOCATION OF BOUNDARY DEFINITION OF BURIED TRENCHES	X	X	(X)	(X)	(X)	(X)
LOCATION AND DEFINITION OF BURIED METALLIC OBJECTS (E.G., DRUMS, ORDINANCE)	(X)	(X)			X	X

X - PRIMARY METHOD
(X) - SECONDARY METHOD

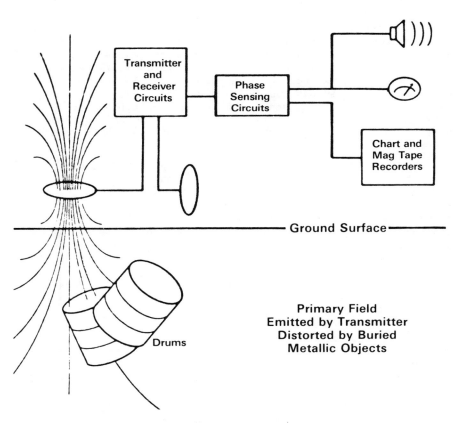

Figure 1.   Simplified diagram of metal detector system.

device. Such a device consists of a transmitter loop coil and a
receiver loop coil geometrically arranged in such a way that a
null or zero-induction condition is achieved in the absence of
nearby metal objects. When the coils are properly adjusted, the
receiver is insensitive to the electromagnetic field of the
transmitter. However, a signal will be detected if the
electromagnetic field at the receiver coil is altered by the
superposition of a secondary electromagnetic field resulting
from currents induced in a nearby metallic object or conductive
mass by the transmitter coil. Deviations from the balance
condition are detected as electrical signals in the receiver and
are electronically converted to a meter deflection as well as to
an audible tone. Metal detectors tend to be insensitive to
buried objects of small cross-section, such as metal rods. (1)
     Figure 2 shows profiles taken with a metal detector at a
rural site where it was suspected that drums containing dioxin
had been dumped into a trench and buried. Multiple parallel
passes were made perpendicular to the suspected trench as part
of an effort to determine the location and quantity of drums in
the trench. (2)

     Magnetometers. Magnetometers can be used to detect
perturbations in the geomagnetic field created by buried
ferromagnetic objects such as steel containers or drums, tools,
or scrap metal. An induced magnetization is produced in any
magnetic material within the earth's magnetic field, and this
induced field is superimposed on the geomagnetic field. If
strong enough, this induced field produces a localized anomaly
in the geomagnetic field. Figure 3 is a schematic of a simple
magnetometer.
     The induced magnetic field of a buried object depends on
several variables:
     1)   Target mass;
     2)   Target to sensor distance;
     3)   Target material and its integrity;
     4)   Target geometry (primarily length to diameter ratio)
     5)   Target orientation;
     6)   Magnitude and direction of permanent magnetism in the
          target
     A variety of magnetometers are currently available for
different types of surveys. Both total field and gradiometer
search magnetometers are used. Proton precession and cesium
systems are used for measurements of the total magnetic field.
Fluxgate and cesium gradiometers can be used for search work. A
gradiometer is a differential magnetometer which measures
differences in the magnitude and direction of the ambient field
over a fixed distance; this fixed distance is usually small with
respect to the distance to the object creating the magnetic
anomaly. An advantage of some gradiometer systems is their
ability to sense vertical field gradients while remaining
insensitive to horizontal gradient components. This feature

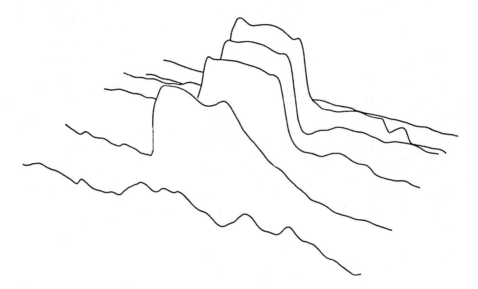

Figure 2.   Metal detector data over trench with buried drums.

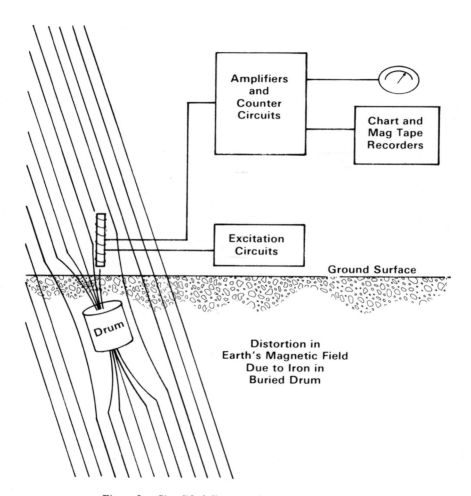

Figure 3.   Simplified diagram of magnetometer system.

allows gradiometers to sense the vertical field of subsurface targets in the presence of horizontal interference targets such as steel fences. (3)

Figure 4 shows multiple parallel profiles taken with a continuous reading gradiometer over the drum-filled burial trench (the same trench from which Figure 2 was taken). These profiles, which add confirmation of the presence of drums in the trench, were taken inside a high steel fence erected around the suspected burial site. (2)

Ground-Penetrating Radar. The ground-penetrating radar (GPR) system most frequently used in hazardous waste site investigations is an impulse system which radiates short-duration electromagnetic pulses into the ground from an antenna near the surface. These pulses are reflected from various interfaces within the earth and are picked up by the receiver section of the antenna and returned to the control unit for processing and display. These reflections occur at different soil horizons, soil/rock interfaces, rock/air interfaces (voids), manmade objects, or at any interface which creates a contrast in complex dielectric properties. For example, digging a trench and filling it again can create a difference between the dielectric properties of the disturbed earth and those of the undisturbed material which can be sensed by the GPR. Figure 5 is a schematic showing the various components of a GPR system.

For presentation of data, GPR signals are processed and displayed by a graphic recorder. As the antenna is moved along the surface, the graphic display results in a picture-like record showing a continuous profile along a traverse, very similar to a geologic cross-section found at a roadcut.

Unfortunately, the depth of radar penetration is very site specific. Depths of 3 to 10 meters are commonly attained throughout the country; 20 meter penetrations have been achieved under ideal conditions at some sites. This depth is reduced if ground water increases in electrical conductivity, or if there are sufficiently high concentrations of fine grained materials (silts or clays) present. For example, high concentrations of salts, montmorillonite clay or losses are highly attenuative of the radar pulse and penetration may not exceed one meter. (4)

Figure 6 shows the cross-section resulting from a single GPR traverse across the barrel-filled trench from which Figures 2 and 4 were taken. The images produced by three 55-gallon barrels are indicated on the figure. Figure 7 is a composite, comparing the traces of Figures 2, 4, and 6. (5)

Resistivity. The resistivity method measures the electrical resistivity of the geohydrologic section, which includes soil, rock, and ground water. Interpretation of these measurements provide information on layering and depths of subsurface horizons as well as lateral changes in the

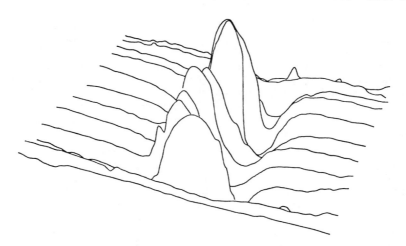

Figure 4.   Magnetometer data over trench with buried drums.

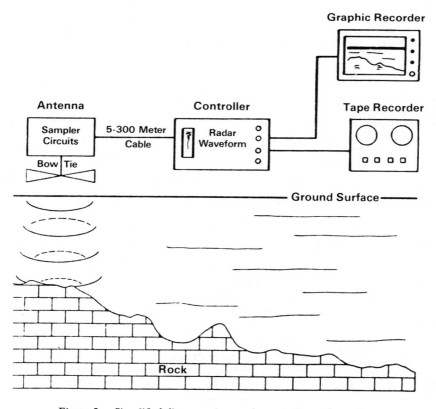

Figure 5.   Simplified diagram of ground penetrating radar system.

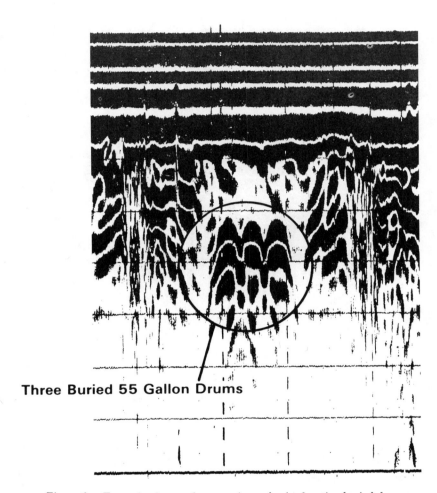

Three Buried 55 Gallon Drums

Figure 6.   Example of ground penetrating radar for locating buried drums.

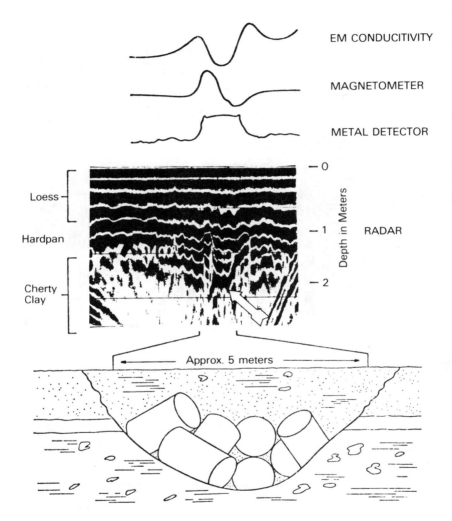

Figure 7.   Composites of four geophysical profiles over hazardous material burial site.

subsurface. In most applications, the presence, quantity and
quality of ground water are the dominating factors influencing
the resistivity value. Accordingly, the method may be used to
assess lateral and vertical changes in natural geohydrologic
settings as well as to evaluate contaminant plumes at hazardous
waste sites.

During application of the method, a current is injected
into the ground by a pair of surface electrodes and a resulting
potential field is measured between a second pair of surface
electrodes. A schematic of such a configuration called a Wenner
array, is shown in Figure 8. The subsurface resistivity is
calculated from the electrode separation, applied current, and
measured voltage.

In general, most soil and rock minerals are electrical
insulators (high resistivity) and, as a result, the flow of
current is conducted primarily through the moisture-filled pore
spaces within this matrix. Therefore, the resistivity of soils
and rocks is predominantly controlled by the amount of pore
water, the porosity and permeability of the system, and the
concentration of dissolved solids in the pore water.

In the field, the resistivity technique may be applied by
"profiling" or "sounding." In profiling horizontal variations
in resistivity are observed by moving the entire array laterally
over the surface. This approach can be used to map the
horizontal extent of contaminant plumes in groundwater.
Soundings are obtained by observing the resistivities which
result from progressively greater electrode spacings, which lead
to correspondingly greater depths of penetration. When these
resistivity values are plotted as a function of electrode
spacing, the resulting plot can be interpreted in terms of the
depths and thicknesses of subsurface layers of differing
resistivities. Figure 9 shows such a sounding in a situation
where alluvial overburden approximately seven feet thick with a
resistivity of approximately 3,000 ohm-feet covers limestone
with a resistivity of approximately 350 ohm-feet. The two
layers are clearly delineated in the resistivity sounding.

Electromagnetics (EM). Electromagnetics (EM) measurements
provide similar data as obtained using the resistivity method;
these data are called ground conductivities (or reciprocal
resistivities). Some of the newer EM techniques are portable,
permitting data to be gathered as fast as a man can walk.
Therefore, the EM method has an advantage over the older
resistivity method in that subsurface conductivities (reciprocal
resistivities) can be collected rapidly and continuously as the
operator and instrument move across the land surface.

The principle of operation of the EM method is shown in
Figure 10. The basic instrument consists of two coils and an
electronics module. The transmitter coil is separated from the
receiver coil by a specified distance. When energized, the
transmitter coil induces circular eddy current loops into the

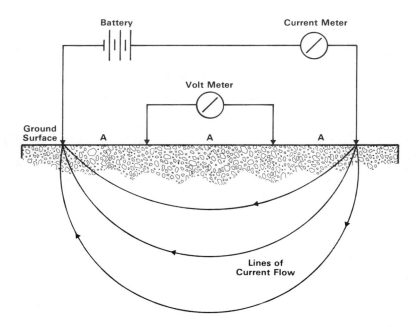

Figure 8.   Simplified diagram of Wenner array for resistivity measurements.

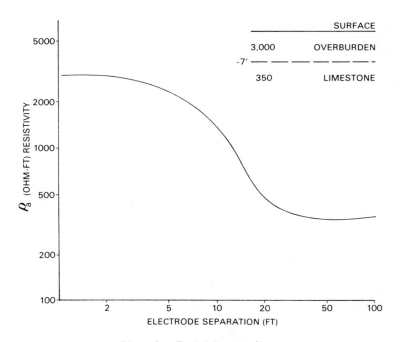

Figure 9.   Resistivity sounding.

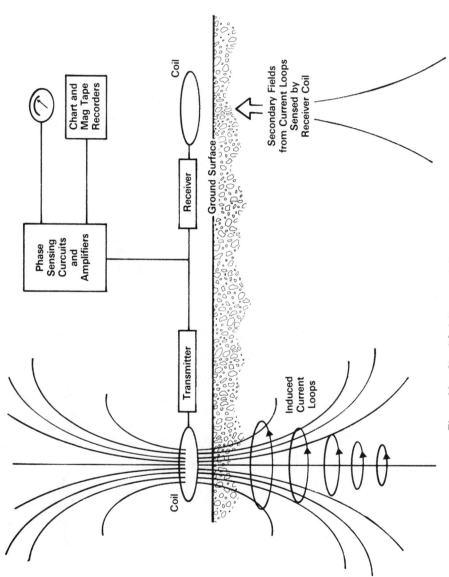

Figure 10.   Simplified diagram of electromagnetic system.

earth. The magnitude of each current loop is a function of
subsurface conditions. In turn, each of these current loops
generates a secondary magnetic field proportional to the value
of the current flowing within the loop. A portion of this
secondary magnetic field is intercepted by the receiver coil and
results in an output voltage which is amplified by the
instrument. The magnitude of this voltage is linearly related
to the terrain (ground) conductivity. The units of conductivity
measurements are millimhos/meter (mm/m). (6)

EM instruments may be calibrated to read the true
subsurface conductivity within a uniform earth. However,
subsurface conditions are rarely uniform. In a layered earth
where each layer has a different conductivity, the reading will
be dependent upon the thickness of the layers, depth of the
layers from the surface, and specific conductivities of each
layer. The resulting conductivity measurement is a complex
function of all these conditions and is called the apparent
conductivity. A strict solution for this function requires some
knowledge of the layer thicknesses, depths and relative
conductivities. However, not all studies need become this
involved because first order evaluations may be made by noting
the relative lateral changes along a traverse (profiling).
Since data can be obtained as fast as the operator can walk,
this profiling capability is a very powerful EM technique
permitting large areas to be rapidly surveyed yielding
continuous profile lines. These profile data may be recorded on
a strip chart and/or magnetic tape recorder. This qualitative
method can often describe the location, shape and/or the
periodicity of a feature whether it is a clay deposit or a
series of fracture zones. Armed with this type of data, along
with some good ground truth information including knowledge of
the local geology, an evaluation of the profile(s) can be made.
This analysis results in a more complete understanding of
specific features and the overall setting of the site.

EM conductivity data may be obtained at depths of 0.75 to
60 meters, depending on the coil spacing and system
configuration. (6)

Although the EM technique is well suited for profiling,
acquisition of vertical changes in subsurface conditions (called
sounding) can be accomplished. However, sounding with EM is
somewhat limited compared to the resistivity sounding methods.
This is due to the fact that only a limited number of discrete
depths can be measured using the EM method; this is unlike the
variable electrode spacings possible with the resistivity
technique. Although the resistivity method may yield more data,
it requires more time to make the measurements.

The principal value of the EM conductivity method is that
it provides continuous, high resolution data in an extremely
economical manner. This permits reconnaissance investigations
to be performed rapidly and effectively to define the location
and extent of problem areas.

Figure 11 shows EM profiles from a hazardous waste site
investigation over a 25-acre area, together with the locations
of monitoring wells which were installed without benefit of the
geophysical measurement data. The monitoring well locations
were chosen by an experienced hydrogeologist on the basis of his
professional judgement and limited information concerning the
history of the site and its contents. None of the wells thus
installed intersected the areas of high conductivity identified
by the EM survey, although two wells did show low levels of
groundwater contamination. In this instance, prior to
geophysical surveys would have enabled a more efficient
monitoring well design to be performed and could have reduced
the total number of wells installed. (7)

Seismic Refraction

Seismic refraction techniques can measure the density,
thickness, and depth of geologic layers using sound (acoustic)
waves transmitted into the subsurface. These sound waves travel
at different velocities in various soils and rock and are also
refracted (or bent) at the interface between layers, thereby
affecting their path of travel. The time required for the wave
to complete this path is measured, permitting determination of
the number of layers at the site as well as the sound velocity
and depth of each layer. The wave velocity in each layer is
related to layer properties such as density and hardness.
Seismic methods are often used to map depth to specific horizons
such as bedrock, clay layers and water table.

Several system components are fundamental to seismic
measurements: a sound source, geophones, and a seismograph.
The seismic source in shallow hazardous waste site
investigations is generally a hammer or drop weight which is
used to strike the ground. Geophones are receivers implanted on
the ground's surface; they translate the received vibrations of
sound energy to an electrical signal. This signal is displayed
on the seismograph permitting measurement of the arrival time of
the sound wave.

The seismic refraction method is based on several important
assumptions: (1) layer acoustic velocities increase with depth;
(2) sufficient velocity contrast exists between layers to
discriminate between different strata of interests; (3) and
layers must be thick enough to permit detection.

Seismic refraction can be used to define many natural and
geohydrologic conditions including number and thickness of
layers, layer composition and physical properties, depth to
bedrock or water table, and anamalous feature.

Several different types of sound energy (waves) are
propagated through the earth. Seismic refraction methods are
concerned primarily with the compressional wave energy, commonly
called primary wave or P-wave. Primary waves move through

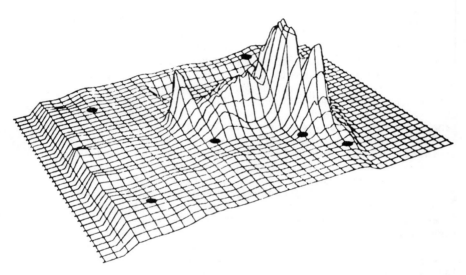

Figure 11.   Three-dimensional representation of conductivity data showing buried hazardous materials. Key: ●, well.

subsurface geologic layers in response to layer physical
properties, thickness and sequence. A significant change in any
one of these parameters will cause a notable shift in the sound
wave's velocity and path of travel which may be detected using
the seismic method.

Layer density and elastic properties primarily determine
the speed or velocity at which sound will travel through the
layer. Porosity, mineral composition, and water content of the
layer affect density and elasticity.

Figure 12 shows a schematic view of sound waves traveling
through a two layered system of soil over bedrock. A seismic
source produces sound waves which travel in all directions into
the ground. The seismic method is only concerned with the wave
paths shown in Figure 12. One of these waves, the direct wave,
travels parallel to the ground's surface in the surface layer.
A seismic sensor (geophone) detects the direct wave as it moves
along the surface layer. The time of travel along this path is
directly proportional to the sensors' distance from the source
and the material composing the layer.

Risk Assessment in Hazardous Waste Site Investigations

Monitoring to estimate exposure to environmental
contaminants is an essential part of assessing risks associated
with those contaminants. Environmental samples are collected in
various media near a source of exposure--for example, a
hazardous waste site suspected of releasing toxic chemicals to
the environment. These samples are then analyzed for the
substances of concern, yielding the basis for estimates of
exposure. The discussion above outlined ways in which
geophysical sensing techniques can be used to collect samples of
groundwater in an efficient, cost-effective manner and to
improve the likelihood of discovering leakage from the hazardous
waste site, if any. Given concentrations of toxic chemicals in
air, water, groundwater, or other exposure medium, risk
assessment can proceed by any or all of several approaches.

In classical epidemiological studies, risks associated with
exposure to toxic chemicals or other environmental contaminants
are assessed by comparing incidences of disease observed in a
population exposed to the etiologic agent with incidences of the
same disease observed in a control population. Relative risk is
then computed as the ratio of the disease rate of the exposed
population to that of the unexposed population. (8) The full
machinery of classical epidemiology is quite expensive and
lengthy, especially the carefully constructed prospective
studies which yield the most reliable data. A more commonly
used approach to risk assessment of specific sources of environmental contaminants, such as hazardous waste sites, is to
compare observed concentrations of the contaminant with existing
standards for that substance in air, water, or food. An
approach which has been suggested for groundwater, perhaps the

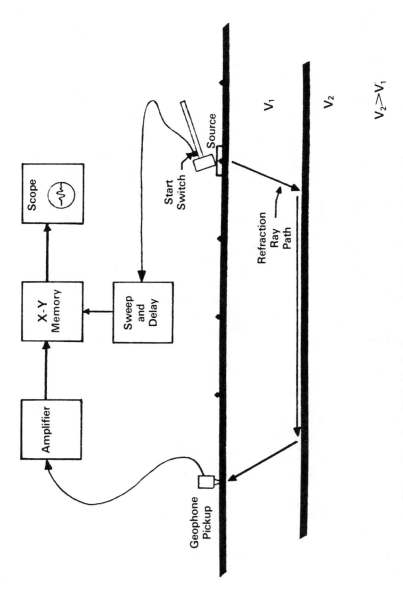

Figure 12. Simplified block diagram of single-trace seismic refraction system.

most important exposure medium in hazardous waste site
investigations, is the use of Maximum Contaminant Levels (MCL's)
for specific chemicals in drinking water. Unfortunately,
hundreds to thousands of chemical species may enter groundwater
from improperly constructed hazardous waste sites, and only a
few of them have established MCL's. EPA is developing
additional MCL's for organic contaminants, but several years
will elapse before the list is comprehensive.

Another approach to risk assessment which does not yield a
quantitative estimate of relative risk in terms of disease
rates, is through comparison of concentrations of the toxic
agent in environmental media near the source with concentrations
in the same media at some control location distant from the
source. Comparisons can also be performed against nationwide
averages. Such comparisons can only establish whether statistically significant differences exist between concentrations in
the area of concern and the national background level. The
comparisons provide no estimate of the biological risk
associated with any differences. All three of these approaches
to risk assessment (the classical epidemiological approach, the
short-cut comparison of observed concentrations with previously
established standards, and the qualitative comparison of
concentrations in the affected area with those in some control
area or with national norms) depend on valid, representative
monitoring data to estimate exposures to populations at risk.

Groundwater contamination by toxic chemicals has recently
become a serious concern to the environmental community; one
recent comparison of concentrations of 56 toxic chemicals in
samples of groundwater with those in samples of surface water in
New Jersey concluded that, at least in New Jersey, groundwater
is at least as contaminated as surface water. (9) Most of the
115 uncontrolled hazardous waste sites identified as highest
priority for cleanup under CERCLA involve groundwater
contamination. (10) Even preliminary assessment of risks at
hazardous waste sites involves determining whether or not
groundwater contamination has occurred, and if so, whether
drinking water supplies are at risk. Definition of the
boundaries of the contaminant plume is a necessary step in
estimating the risk to drinking water supplies. Groundwater
monitoring is essential to assessing exposure through drinking
water at locations where wells are important as drinking water
supplies.

Summary

Currently available geophysical methods most applicable in
hazardous waste site investigations include: metal detectors,
magnetometers, ground-penetrating radar (GPR), electromagnetic
induction (EM), resistivity, and seismic refraction. These
methods should be regarded as complementary, since no one method

will satisfy all requirements, and some of these are useful in
more than one application. In general, metal detectors and
magnetometers are useful in locating buried wastes; GPR is
useful in defining the boundaries of buried trenches; EM is most
useful in defining the horizontal extent of plumes of conductive
contaminants in groundwater; and resistivity and seismic
refraction are most useful in determining geologic stratigraphy.
A cost-effective approach to hazardous waste site assessments
would include three phases: (1) preliminary site assessment,
involving the use of aerial photography, on-site inspections,
and readily available information to approximate site boundaries
and locations of buried wastes, as well as probable site
geology; (2) geophysical surveys to pinpoint buried waste
locations and to estimate quantities; and (3) confirmation of
groundwater contamination through monitoring networks designed
on the basis of contaminant plumes and stratigraphy revealed by
the geophysical survey. Spatial characterization of the site by
geophysical methods can make possible the efficient location of
monitoring wells and the reduction of risks involved in
exploratory surveys.

Acknowledgement

      The author gratefully acknowledges the permission of Mr.
Richard Benson of Technos, Inc. to use the data and figures
presented here.

Literature Cited

1. Sandness, G. A.; Dawson, G. W.; Mathieu, T. J., and
      Rising, J. L. "The Application of Geophysical Survey
      Techniques to Mapping of Wastes in Abandoned Landfills,"
      Proceedings of the 1979 Conference on Hazardous Materials
      Risk Assessment, Disposal and Management, April 25-27,
      1979, Miami Beach, Florida.
2. Benson, R. Technos, Inc. Personal Communication, 1982.
3. Breiner, S. "Applications Manual for Portable
      Magnetometers," GeoMetrics, Sunnyvale, California, 1973.
4. Johnson, R. W.; Glaccum, R.; Wojtasinski. "Application of
      Ground Penetrating Radar to Soil Surveys," Proceedings of
      the Soil and Crop Science Society of Florida, Vol. 39,
      1980.
5. Benson, R. C.; Glaccum, R. A. "Site Assessment: Improv-
      ing Confidence Levels with Surface Remote Sensing,"
      Proceedings of the U.S. EPA National Conference on
      Management of Uncontrolled Hazardous Waste Sites, October
      15-17, 1980, Washington, D.C.
6. McNeill, J. D. "Electromagnetic Terrain Conductivity
      Measurement at Low Induction Numbers," Technical Note
      TN-6, Geonics Ltd., October 1980.

7.   Benson, Richard; Glaccum, Robert; Beam, Paul. "Minimizing
     Cost and Risk in Hazardous Waste Site Investigations Using
     Geophysics," in proceedings of the National Conference on
     Management of Uncontrolled Hazardous Waste Sites, October
     28–30, 1981, Washington, D.C.
8.   Lilienfeld, Abraham M. "Foundations of Epidemiology,"
     Oxford University Press, New York, 1976, p. 179.
9.   Page, G. William. "Comparison of Groundwater and Surface
     Water for Patterns and Levels of Contamination by Toxic
     Substances," Environmental Science and Technology, Vol.
     15, No. 12, December 1981.
10. "Interim List of Superfund Priority Waste Sites," in
     Hazardous Materials Intelligence Report, October 23, 1981.

RECEIVED August 3, 1982.

Incorporating Risk Assessment into the Resource Conservation and Recovery Act Regulatory Process

CURTIS HAYMORE

U.S. Environmental Protection Agency, Office of Solid Waste, Washington, DC 20460

This paper reviews the types of risk assessments being conducted to develop regulations for the control of hazardous wastes under the Resource Conservation and Recovery Act. Better data and Executive Order 12291 have spurred the development of more detailed risk assessment models in our regulatory analyses. The first of three projects is a broad policy overview of the entire hazardous waste program that focuses on the risks and costs of different combinations of wastes, technologies, and environmental settings. The second project is a series of Regulatory Impact Analyses that will review our first wave of several regulations. The third project is designed to tailor regulations to specific industries and waste streams.

The regulation of hazardous waste is one of the most complex problems with which the Environmental Protection Agency has had to deal. Hazardous wastes generally contain many substances that adversely affect human health and the environment. The types of effects differ; the timing of their occurrence varies; the waste streams change through time, and often no two batches are the same, even when they are from the same plant that runs the same process continuously. And the wastes come from firms in all industries, large and small, profitable and marginal. This complexity has prevented EPA from regulating hazardous waste activities as quickly and as thoroughly as we and Congress had hoped. We will, however, soon have a full set of regulations that the States and EPA can implement. We are now reassessing those regulations and hope to propose any necessary revisions in 1983.

One of the most important goals of the Resource Conservation and Recovery Act of 1976 was to protect human health and the environment from the risks posed by hazardous waste. The measurement of those risks and the determination of how best to reduce them involve assessing the inherent hazards of different

0097-6156/82/0204-0117$06.00/0

chemicals, their potential for exposure, and the consequent effects of various waste streams and their chemical constituents
on people and the environment. Analyzing these risks, even if
in a very crude way, is absolutely essential to the development
of our regulatory program. We cannot determine what we want our
regulations to accomplish, we cannot compare alternative methods
of achieving our goals, and we cannot gauge our success without
some form of risk assessment. Even though there is considerable
uncertainty involved in this area, we in the Office of Solid
Waste are deeply committed to using such assessments to improve
our ability to produce fair and effective regulations.

     We hope to use these types of analyses in making all of our
major decisions. Although in many areas we will only be able to
analyze gross differences in risk, we feel that some information
is better than none. Comprehensive risk assessments are our
goal, but the precision of our efforts will vary greatly in different projects depending on the data and predictive models
available.

     In the past, the Office of Solid Waste did not explicitly
use formal risk assessments in developing regulations. The regulations for hazardous waste facilities proposed by EPA in 1978
were largely design and performance standards. We did this because we had very limited data, and there was no recognized state
of the art. In 1981, we incorporated risk assessment into what
was called the best engineering judgment (or BEJ) approach to regulating the land disposal and incineration of hazardous waste.
These assessments were to be done on a site-by-site basis, with
the owners and operators responsible for demonstrating that their
facilities would not pose a threat to human health and the environment. Some members of the regulated community objected to the
uncertainty involved in that approach. They felt requirements
would vary in unpredictable ways for each site. EPA now agrees
that we should provide better guidance about what levels of control are acceptable. EPA's new regulations for land disposal reflect this need for better guidance and will primarily use environmental performance standards. If facilities cannot meet those
standards, they can still attempt to make a demonstration based
on a risk assessment that their operation is acceptably safe even
if the performance standard is violated.

     In the reexamination of our existing regulations and forth-
coming land disposal regulations, the Office of Solid Waste is
using risk assessments as a part of its internal review. We believe, first of all, that as much analysis as possible should be
done beforehand and be incorporated into the regulations to minimize costs and promote consistency. Second, we now have and are
continuing to obtain much better data. Such information, which
was not available to the Agency prior to promulgation of the current regulations, will contribute significantly to our knowledge
of the number, types, and locations of hazardous waste management
facilities and the wastes they handle, and, consequently, to

meaningful risk analysis. A third reason for using more risk assessment in our development and review of regulations is the friendly encouragement of Executive Order 12291, which requires that cost-benefit studies be completed on all major regulations to the greatest extent possible.

We must recognize, however, that our abilities may be limited by a lack of other types of data and by the limitations of the rapidly evolving science of risk assessment. In an effort to minimize these limitations, the Office of Solid Waste is investigating the best available risk assessment techniques. These include estimation of the movement of pollutants through soil, air, and water; prediction of adverse human health and environmental effects on the basis of available toxicity data; and prediction of the effects of simultaneous exposures to numerous toxic substances. OSW is, in addition, actively compiling data relative to the cost, applicability, and effectiveness of currently available waste treatment, storage, and disposal technologies.

Specific work we now have under way falls into three categories: our RCRA strategy project, a series of Regulatory Impact Analyses, and detailed studies on specific industries and waste streams. The strategy project is our broadest approach to evaluating risks and costs. It is an attempt to achieve an overview of our program and to determine our general constraints. What are the approximate costs to society for different degrees of environmental safety? What happens when wastes are controlled to different levels by means of different technologies? To what extent do wastes flow from safer technologies to less safe technologies because of lower costs of disposal and treatment? We are trying to answer these questions by looking at waste management situations in a special way. We ask what wastes we have, what we are doing with them (that is, what type of technology we are using and to what level of control), and where we are physically in terms of the environmental setting or location. These combinations of wastes, environments, and technologies constitute our unit of analysis. Some people call this the W E T or WET approach. It allows us to compare the degree of hazard posed by different waste management situations.

Our project differs from other degree of hazard approaches in two important respects. First, we are not simply trying to take a list of wastes and waste streams and divide them into categories of intrinsic hazard. We are attempting, in general, to consider their potential for effects on humans, and that requires a careful consideration of technologies and environmental settings, as well as inherent hazard. What may result from this approach are groupings of wastes whose hazards vary according to the environmental setting or the technology used.

The second important advantage over other degree of hazard approaches is that we are assessing not just risk. We are, in addition, trying to assess the costs of achieving the various levels of control. This combination of the amount of environmental and health safety we can buy with varying amounts of society's resources is the important and relevant policy issue.

How can we do this assessment for the entire range of hazardous waste situations? We obviously have had to make many simplifying assumptions. First, we are initially limiting our review to human health effects. Second, we treat risk very broadly -- our system only distinguishes between tenfold differences in risk. Third, only direct capital and operating expenses are estimated. Fourth, we are assuming that different chronic health effects can be aggregated into a single value by scoring the probability of incidences per unit dose. And, finally, we are initially ignoring acute effects such as those from fires and explosions.

Let me spend a minute describing our methodology. We specify 22 treatment technologies and 9 disposal technologies. Our treatment technologies include such things as vacuum filters, chemical precipitation, and four levels of incineration. Disposal technologies include three levels of landfills (distinguished by the degree of stringency), one type of underground injection well, three types of surface impoundments, and so forth. For each technology, we specify the types of waste streams it can and cannot handle. Waste streams composed of primarily heavy metals, for example, will not be incinerated. We specify what effect the treatment steps have on the waste streams in terms of volumes, concentrations of hazardous constituents, and so forth. We then estimate what portion of the waste stream entering each technology, both treatment and disposal, escapes into the different environmental media -- air, surface water, and ground water. For each volume of waste going through the model, the releases to each media are summed. We also compute the cost of using this string or chain of treatment and disposal technologies, including high, low, and typical costs. Costs, in this model, are direct capital and operating expenses and exclude transfer payments such as taxes and fees. We express these costs as a score, where each higher value represents roughly a doubling of cost per unit of waste. At this point, we have release rates of waste streams and their chemical components from different combinations of technologies and the associated costs. Now we must determine how dangerous these releases are. To do so, we switch to the risk side of our model.

To establish risk, we have initially scored 140 chemical compounds on what we are calling the inherent hazard of the compound. For the first round of the project, we are only looking at the risk to human health and are excluding economic and ecological risks. Our goal when scoring for risk to human health is to have a single value represent all types of effects from different types of cancers to different kinds of graded responses such as liver damage. The conceptual link we are trying to use is, as I mentioned earlier, the probability of an incident per unit dose. A score of "2" on our scale, for example, is intended

to represent roughly a 1 percent risk of either contracting cancer o̲r̲ having an adverse effect from consuming 1 mg of pollutant
for every 1 kg of body weight per day. We obviously make some
gross assumptions to allow our conceptual model to use published
data. The model assumes, specifically, a linear dose response
for carcinogens. It also assumes that most reported MEDs, or
minimum effective doses, correspond to a risk of about 10 percent. We will try to test the sensitivity of our results in re-
lation to these assumptions as the project continues.

The system then adjusts these scores to account for the way
different compounds react in the different media -- air, ground
water, and surface water. On the basis primarily of a compound's
half-life in a medium and on dispersion patterns, we assign each
compound a separate inherent risk score for each medium. The
scale we use is very coarse: each level is 10 times greater than
the previous level. The data now in the model is, therefore,
insensitive to risks that are only two or thre times as great as
others. We found it convenient to express the ten-fold differences on a logarithmic unit scale.

We next attempt to calibrate the model, or explain what a
score of "8," for instance, means in our system in terms of releases to the environment. We conclude with a series of release
rates for compounds into different media that reflect essentially
the same risk. A score of "8," for example, may mean a release
of 32 tons per year to air of copper or 63 tons per year of
nickel to surface water.

Let us quickly review. On the technology side, we develop
the release rates of different waste streams into different media. On the waste side, we have a way of expressing and comparing how hazardous these releases of the chemical components in
the waste stream are. After making some adjustments for different environmental settings, we can score the risk and correspond-
ing cost of disposing waste streams using different technologies
in different places. By putting constraints on the volumes of
waste streams and the capacities of different technologies, we
can then use a linear programming model to achieve some objective
function, or goal. We are trying several different objective
functions, but we will in general look at the cost of achieving
various levels of risk.

The work is being done under contract by ICF, Inc., SCS
Engineers, and Clement Associates. A computer model incorporating these assumptions should be built and documented by May.
We will then be adjusting, testing, and expanding the model to
improve its performance. We intend to be able to use the model,
even in the short run, for a number of purposes. We can perhaps
identify combinations of wastes, environments, and technologies
where regulatory control can decrease or increase or where, at
the extreme, we can prohibit certain practices or totally eliminate regulatory requirements. We can identify those alternatives
that seem most promising for detailed cost-benefit analyses. The

RCRA strategy project is -- it must be stressed -- a broad policy planning tool; it does not, and is incapable of, developing and revising specific regulations.

The second category of risk assessments being done by OSW is part of our largest analytical effort; a series of Regulatory Impact Analyses (RIAs) that we are conducting as required by Executive Order 12291. These cost-benefit analyses are the real basis for any regulatory revisions we will make in the future. Most of the RIAs focus on specific technologies -- storage facilities, landfills, incinerators, surface impoundments, and land treatment facilities. Two are related to environmental settings: seismic areas and floodplains. One, waste oil, is focused on a particular waste stream. The RIAs consist, in general, of three steps -- identifying the problem, selecting promising alternatives to solve the problem, and performing the detailed cost-benefit analyses needed to select appropriate regulatory requirements.

Identifying the extent of the problem requires extensive data. We are now collecting these data through about 300 site visits and about 3,000 questionnaries to hazardous waste facilities and generators. The results of these site visits and questionnaires, along with our other technical, economic, and policy studies, should provide the necessary information on which we can base future action.

Some of the RIAs will be more sophisticated than others. The RIA on incineration, for example, will be able to model and, to some extent, quantify the relative risk to health using actual incinerator data, at least for a few chemicals and waste streams, for the most exposed individual and for the entire exposed population. The RIA on incineration uses the Industrial Source Complex dispersion model developed by EPA's Office of Air Quality Planning and Standards. We have or are receiving inhalation toxicity data from the Environmental Criteria and Assessment Office in EPA for most of the compounds listed in Appendix VIII of our regulations. We will run the model assuming different destruction efficiencies, different stack heights, different waste streams, and different population distributions. The model calculates the risk to people in the "hot spots," that is the most exposed individuals, in terms of lifetime risk of cancer. (We are primarily estimating health risks from cancer at this point.) The model also calculates the numbers of probable cancers for the community and region as a whole. We believe that linking this dispersion model to the toxicity estimates yields state of the art results in quantifying the health hazards (or health benefits in terms of avoided health effects) associated with incinerators.

The RIA on. land disposal, conversely, will probably not be as quantitative or specific about the risks posed. In both incineration and land disposal, however, we believe we will be using the state of the art in assessing risks for hazardous waste facilities. These risk assessments are under way, and we expect results sometime next year. You must remember though, that the purpose of these studies is to help develop regulations. The methods used in developing the RIAs are not, in general, designed for assessing the actual risks of specific, individual sites.

The third category of activities we are pursuing in risk assessment is the detailed analysis of specific industries or waste streams. This industry studies program will result in tailored regulations, based on specific waste stream analyses, identified industrial processes, and particular locations. We may either increase or decrease the stringency of current controls or classify new waste streams as hazarous on the basis of our findings. We are now examining chlorinated organic chemicals and organic chemical products such as pesticides, dyes, and pigments. We will soon expand our effort to cover other aspects of the organic chemicals industry. The data gathered under the waste characterization component will determine the waste streams of concern from specific production processes for listing as hazardous wastes. The waste management component will also help in listing wastes by characterizing existing practices. In addition, the approach will provide decisionmakers with information that they can use in tailoring management standards.

How successful will we be in all our efforts? I believe we will greatly advance our understanding of the activities we are regulating, and, as a result, we are very likely to be able to improve the efficiency of our regulations. We should be able to obtain more safety, at a lower cost, with fewer burdens and misplaced incentives to industry, and with more consistency than we do now. Our RCRA strategy project should enable us to avoid big mistakes and should provide a basis for making better decisions than in the past.

Our assessment of the risks posed by incinerators should be the most structured of our Regulatory Impact Analyses. The detail we will be able to achieve in the other areas is uncertain, but we believe our work will represent the best that can be accomplished now. We need better tools to analyze the effects of our programs. Our two primary tools are economic analysis and risk analysis. We plan to rely heavily on them. To do less would be short-sighted and irresponsible.

RECEIVED June 16, 1982.

INDEX

125